Easy Algebra
STEP-BY-STEP

Master High-Frequency Concepts and Skills for Algebra Proficiency—*FAST!*

Sandra Luna McCune, Ph.D., and William D. Clark, Ph.D.

New York Chicago San Francisco Lisbon London Madrid Mexico City
Milan New Delhi San Juan Seoul Singapore Sydney Toronto

The McGraw-Hill Companies

1 2 3 4 5 6 7 8 9 10 11 12 13 14 15 QFR/QFR 1 9 8 7 6 5 4 3 2 1

ISBN 978-0-07-176724-8
MHID 0-07-176724-X

e-ISBN 978-0-07-176725-5
e-MHID 0-07-176725-8

Library of Congress Control Number 2011928668

Other titles in the series:
Easy Biology Step-by-Step, Nichole Vivion
Easy Chemistry Step-by-Step, Marian DeWane
Easy Mathematics Step-by-Step, Sandra Luna McCune and William D. Clark
Easy Precalculus Step-by-Step, Carolyn Wheater

This book is printed on acid-free paper.

Contents

Preface vii

1 **Numbers of Algebra 1**

Natural Numbers, Whole Numbers, and Integers 1
Rational, Irrational, and Real Numbers 4
Properties of the Real Numbers 8

2 **Computation with Real Numbers 14**

Comparing Numbers and Absolute Value 14
Addition and Subtraction of Signed Numbers 17
Multiplication and Division of Signed Numbers 26

3 **Roots and Radicals 32**

Squares, Square Roots, and Perfect Squares 32
Cube Roots and *n*th Roots 36
Simplifying Radicals 39

4 **Exponentiation 44**

Exponents 44
Natural Number Exponents 45
Zero and Negative Integer Exponents 48
Unit Fraction and Rational Exponents 53

5 **Order of Operations 58**

Grouping Symbols 58
PEMDAS 60

6 Algebraic Expressions 64
Algebraic Terminology 64
Evaluating Algebraic Expressions 66
Dealing with Parentheses 70

7 Rules for Exponents 74
Product Rule 74
Quotient Rule 75
Rules for Powers 77
Rules for Exponents Summary 79

8 Adding and Subtracting Polynomials 83
Terms and Monomials 83
Polynomials 85
Like Terms 87
Addition and Subtraction of Monomials 88
Combining Like Terms 89
Addition and Subtraction of Polynomials 90

9 Multiplying Polynomials 94
Multiplying Monomials 94
Multiplying Polynomials by Monomials 96
Multiplying Binomials 98
The FOIL Method 99
Multiplying Polynomials 101
Special Products 102

10 Simplifying Polynomial Expressions 104
Identifying Polynomials 104
Simplifying Polynomials 106

11 Dividing Polynomials 110
Dividing a Polynomial by a Monomial 110
Dividing a Polynomial by a Polynomial 113

12 Factoring Polynomials 119
Factoring and Its Objectives 119
Greatest Common Factor 120
GCF with a Negative Coefficient 123
A Quantity as a Common Factor 125
Factoring Four Terms 126

Factoring Quadratic Trinomials 127
Perfect Trinomial Squares 133
Factoring Two Terms 134
Guidelines for Factoring 137

13 Rational Expressions 139

Reducing Algebraic Fractions to Lowest Terms 139
Multiplying Algebraic Fractions 143
Dividing Algebraic Fractions 145
Adding (or Subtracting) Algebraic Fractions, Like Denominators 146
Adding (or Subtracting) Algebraic Fractions, Unlike Denominators 148
Complex Fractions 151

14 Solving Linear Equations and Inequalities 154

Solving One-Variable Linear Equations 154
Solving Two-Variable Linear Equations for a Specific Variable 159
Solving Linear Inequalities 159

15 Solving Quadratic Equations 163

Solving Quadratic Equations of the Form $ax^2 + c = 0$ 163
Solving Quadratic Equations by Factoring 165
Solving Quadratic Equations by Completing the Square 166
Solving Quadratic Equations by Using the Quadratic Formula 167

16 The Cartesian Coordinate Plane 171

Definitions for the Plane 171
Ordered Pairs in the Plane 171
Quadrants of the Plane 174
Finding the Distance Between Two Points in the Plane 176
Finding the Midpoint Between Two Points in the Plane 177
Finding the Slope of a Line Through Two Points in the Plane 178
Slopes of Parallel and Perpendicular Lines 181

17 Graphing Linear Equations 184

Properties of a Line 184
Graphing a Linear Equation That Is in Standard Form 184
Graphing a Linear Equation That Is in Slope-y-Intercept Form 186

18 The Equation of a Line 189

Determining the Equation of a Line Given the Slope and y-Intercept 189
Determining the Equation of a Line Given the Slope and One Point on the Line 190
Determining the Equation of a Line Given Two Distinct Points on the Line 192

19 Basic Function Concepts 195

Representations of a Function 195
Terminology of Functions 197
Some Common Functions 201

20 Systems of Equations 205

Solutions to a System of Equations 205
Solving a System of Equations by Substitution 206
Solving a System of Equations by Elimination 208
Solving a System of Equations by Graphing 210

Answer Key 213

Index 239

Preface

Easy Algebra Step-by-Step is an interactive approach to learning basic algebra. It contains completely worked-out sample solutions that are explained in detailed, step-by-step instructions. Moreover, it features guiding principles, cautions against common errors, and offers other helpful advice as "pop-ups" in the margins. The book takes you from number concepts to skills in algebraic manipulation and ends with systems of equations. Concepts are broken into basic components to provide ample practice of fundamental skills.

The anxiety you may feel while trying to succeed in algebra is a real-life phenomenon. Many people experience such a high level of tension when faced with an algebra problem that they simply cannot perform to the best of their abilities. It is possible to overcome this difficulty by building your confidence in your ability to do algebra and by minimizing your fear of making mistakes.

No matter how much it might seem to you that algebra is too hard to master, success will come. Learning algebra requires lots of practice. Most important, it requires a true confidence in yourself and in the fact that, with practice and persistence, you will be able to say, "I can do this!"

In addition to the many worked-out, step-by-step sample problems, this book presents a variety of exercises and levels of difficulty to provide reinforcement of algebraic concepts and skills. After working a set of exercises, use the worked-out solutions to check your understanding of the concepts.

We sincerely hope *Easy Algebra Step-by-Step* will help you acquire greater competence and confidence in using algebra in your future endeavors.

1

Numbers of Algebra

The study of algebra requires that you know the specific names of numbers. In this chapter, you learn about the various sets of numbers that make up the real numbers.

Natural Numbers, Whole Numbers, and Integers

The *natural numbers* (or *counting numbers*) are the numbers in the set

> The three dots indicate that the pattern continues without end.

$$N = \{1,2,3,4,5,6,7,8,\ldots\}$$

You can represent the natural numbers as equally spaced points on a number line, increasing endlessly in the direction of the arrow, as shown in Figure 1.1.

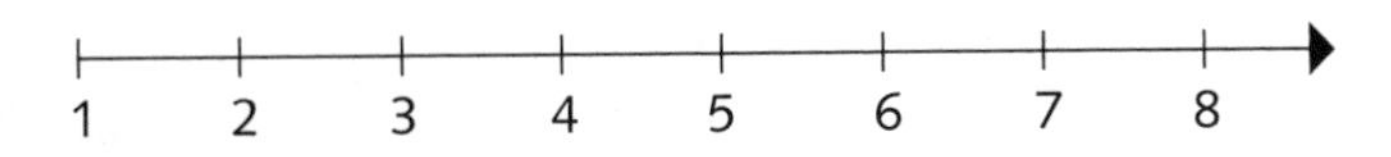

Figure 1.1 Natural numbers

The sum of any two natural numbers is also a natural number. For example, $3+5=8$. Similarly, the product of any two natural numbers is also a natural number. For example, $2\times 5=10$. However, if you subtract or divide two natural numbers, your result is not always a natural number. For instance, $8-5=3$ is a natural number, but $5-8$ is not.

> You do not get a natural number as the answer when you subtract a larger natural number from a smaller natural number.

Likewise, $8 \div 4 = 2$ is a natural number, but $8 \div 3$ is not.

You do not get a natural number as the quotient when you divide natural numbers that do not divide evenly.

When you include the number 0 with the set of natural numbers, you have the set of whole numbers:

The number 0 is a whole number, but not a natural number.

$$W = \{0,1,2,3,4,5,6,7,8,\ldots\}$$

If you add or multiply any two whole numbers, your result is always a whole number, but if you subtract or divide two whole numbers, you are not guaranteed to get a whole number as the answer.

Like the natural numbers, you can represent the whole numbers as equally spaced points on a number line, increasing endlessly in the direction of the arrow, as shown in Figure 1.2.

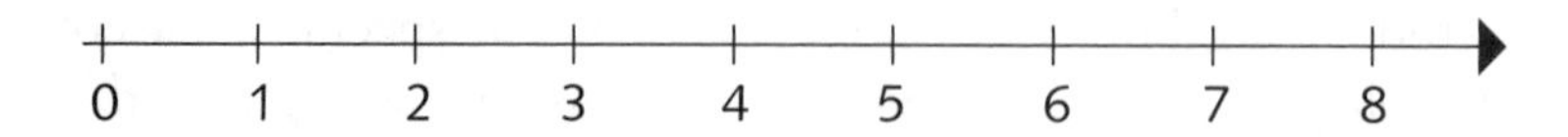

Figure 1.2 Whole numbers

The *graph* of a number is the point on the number line that corresponds to the number, and the number is the *coordinate* of the point. You graph a set of numbers by marking a large dot at each point corresponding to one of the numbers. The graph of the numbers 2, 3, and 7 is shown in Figure 1.3.

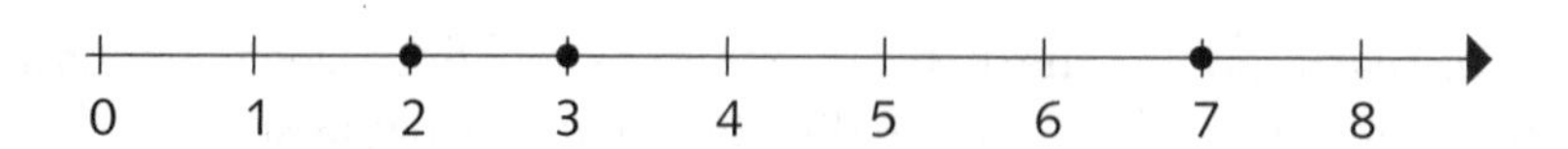

Figure 1.3 Graph of 2, 3, and 7

On the number line shown in Figure 1.4, the point 1 unit to the left of 0 corresponds to the number −1 (read "negative one"), the point 2 units to the left of 0 corresponds to the number −2, the point 3 units to the left of 0 corresponds to the number −3, and so on. The number −1 is the *opposite* of 1, −2 is the opposite of 2, −3 is the opposite of 3, and so on. The number 0 is its own opposite.

A number and its opposite are exactly the same distance from 0. For instance, 3 and −3 are opposites, and each is 3 units from 0.

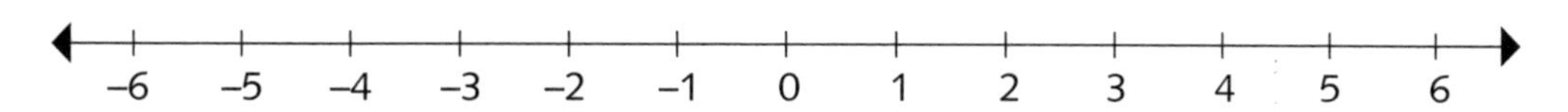

Figure 1.4 Whole numbers and their opposites

The set consisting of the whole numbers and their opposites is the set of integers (usually denoted Z):

$$Z = \{\ldots,-3,-2,-1,0,1,2,3,\ldots\}$$

0 is neither positive nor negative.

The integers are either *positive* (1,2,3,...), *negative* (...,−3,−2,−1), or 0.

Positive numbers are located to the right of 0 on the number line, and negative numbers are to the left of 0, as shown in Figure 1.5.

It is not necessary to write a + sign on positive numbers (although it's not wrong to do so). If no sign is written, then you know the number is positive.

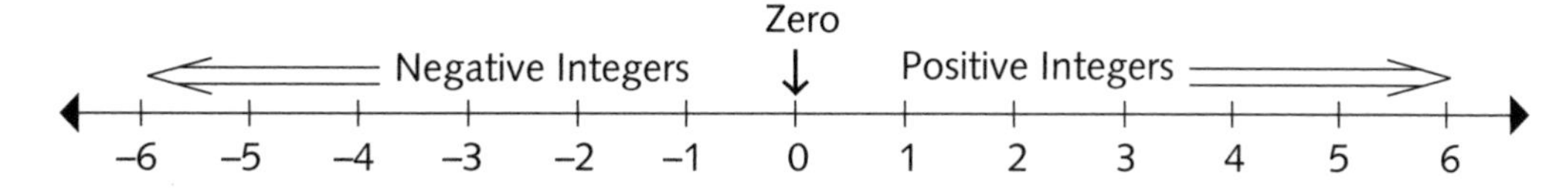

Figure 1.5 Integers

Problem Find the opposite of the given number.

a. 8

b. −4

Solution

a. 8

Step 1. 8 is 8 units to the right of 0. The opposite of 8 is 8 units to the left of 0.

Step 2. The number that is 8 units to the left of 0 is −8. Therefore, −8 is the opposite of 8.

b. −4

Step 1. −4 is 4 units to the left of 0. The opposite of −4 is 4 units to the right of 0.

Step 2. The number that is 4 units to the right of 0 is 4. Therefore, 4 is the opposite of −4.

Problem Graph the integers −5, −2, 3, and 7.

Solution

Step 1. Draw a number line.

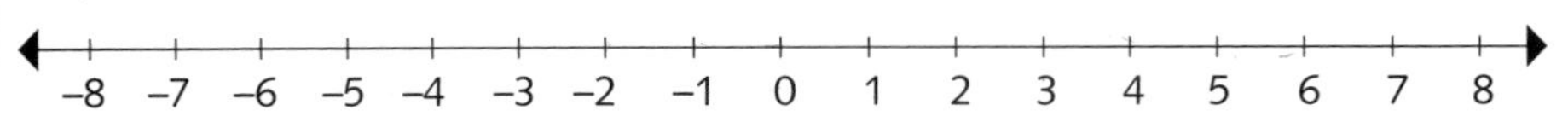

Step 2. Mark a large dot at each of the points corresponding to −5, −2, 3, and 7.

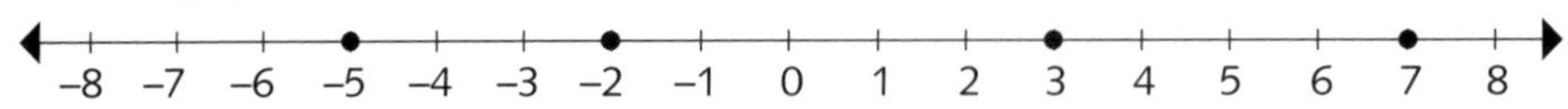

Rational, Irrational, and Real Numbers

You can add, subtract, or multiply any two integers, and your result will always be an integer, but the quotient of two integers is not always an integer. For instance, $6 \div 2 = 3$ is an integer, but $1 \div 4 = \frac{1}{4}$ is not an integer. The number $\frac{1}{4}$ is an example of a rational number.

A *rational number* is a number that can be expressed as a quotient of an integer divided by an integer other than 0. That is, the set of rational numbers (usually denoted Q) is

$$Q = \left\{\frac{p}{q}, \text{ where } p \text{ and } q \text{ are integers, } q \neq 0\right\}$$

> The number 0 is excluded as a denominator for $\frac{p}{q}$ because division by 0 is undefined, so $\frac{p}{0}$ has no meaning, no matter what number you put in the place of p.

Fractions, decimals, and percents are rational numbers. All of the natural numbers, whole numbers, and integers are rational numbers as well because each number n contained in one of these sets can be written as $\frac{n}{1}$, as shown here.

$$\ldots, -3 = \frac{-3}{1}, -2 = \frac{-2}{1}, -1 = \frac{-1}{1}, 0 = \frac{0}{1}, 1 = \frac{1}{1}, 2 = \frac{2}{1}, 3 = \frac{3}{1}, \ldots$$

The decimal representations of rational numbers terminate or repeat. For instance, $\frac{1}{4} = 0.25$ is a rational number whose decimal representation terminates, and $\frac{2}{3} = 0.666\ldots$ is a rational number whose decimal representation repeats. You can show a repeating decimal by placing a line over the block of digits that repeats, like this: $\frac{2}{3} = 0.\overline{6}$. You also might find it convenient to round the repeating decimal to a certain number of decimal places. For instance, rounded to two decimal places, $\frac{2}{3} \approx 0.67$.

The symbol ≈ is used to mean "is approximately equal to."

The *irrational numbers* are the real numbers whose decimal representations neither terminate nor repeat. These numbers cannot be expressed as ratios of two integers. For instance, the positive number that multiplies by itself to give 2 is an irrational number called the positive square root of 2. You use the square root symbol $\left(\sqrt{\ }\right)$ to show the positive square root of 2 like this: $\sqrt{2}$. Every positive number has two square roots: a positive square root and a negative square root. The other square root of 2 is $-\sqrt{2}$. It also is an irrational number. (See Chapter 3 for an additional discussion of square roots.)

The number 0 has only one square root, namely, 0 (which is a rational number). The square roots of negative numbers are not real numbers.

You cannot express $\sqrt{2}$ as the ratio of two integers, nor can you express it precisely in decimal form. Its decimal equivalent continues on and on without a pattern of any kind, so no matter how far you go with decimal places, you can only approximate $\sqrt{2}$. For instance, rounded to three decimal places, $\sqrt{2} \approx 1.414$.

Not all roots are irrational. For instance, $\sqrt{36} = 6$ and $\sqrt[3]{-64} = -4$ are rational numbers.

Do not be misled, however. Even though you cannot determine an exact value for $\sqrt{2}$, it is a number that occurs frequently in the real world. For instance, designers and builders encounter $\sqrt{2}$ as the length of the diagonal of a square that has sides with length of 1 unit, as shown in Figure 1.6.

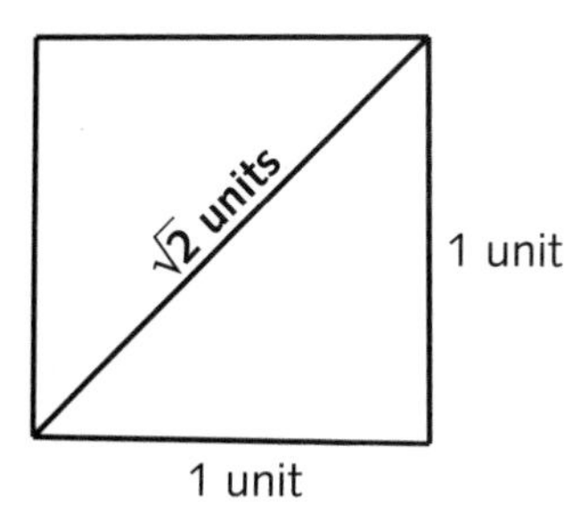

Figure 1.6 Diagonal of unit square

There are infinitely many other roots—square roots, cube roots, fourth roots, and so on—that are irrational. Some examples are $\sqrt{41}$, $\sqrt[3]{-18}$, and $\sqrt[4]{100}$.

> Be careful: Even roots ($\sqrt{\ }$, $\sqrt[4]{\ }$, $\sqrt[6]{\ }$, etc.) of *negative* numbers are not real numbers.

Two famous irrational numbers are π and e. The number π is the ratio of the circumference of a circle to its diameter, and the number e is used extensively in calculus. Most scientific and graphing calculators have π and e keys. To nine decimal place accuracy, $\pi \approx 3.141592654$ and $e \approx 2.718281828$.

> Although, in the past, you might have used 3.14 or $\frac{22}{7}$ for π, π does not equal either of these numbers. The numbers 3.14 and $\frac{22}{7}$ are rational numbers, but π is irrational.

The real numbers, R, are all the rational and irrational numbers put together. They are all the numbers on the number line (see Figure 1.7). Every point on the number line corresponds to a real number, and every real number corresponds to a point on the number line.

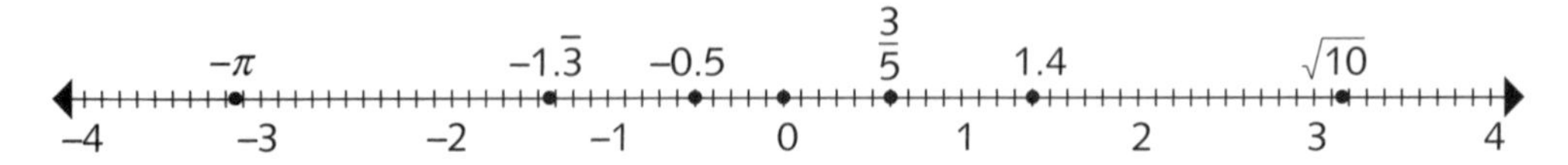

Figure 1.7 Real number line

The relationship among the various sets of numbers included in the real numbers is shown in Figure 1.8.

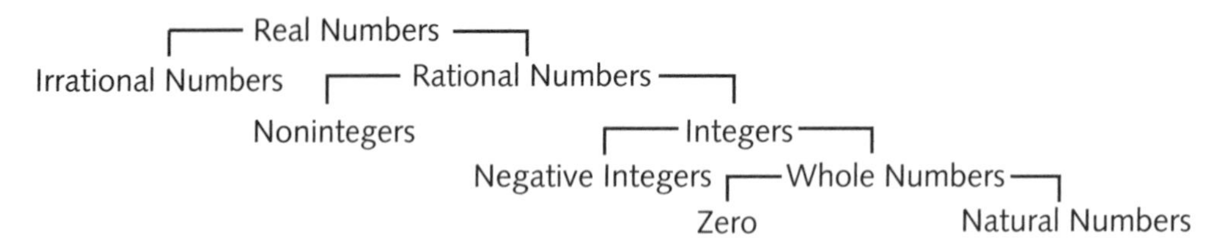

Figure 1.8 Real numbers

Problem Categorize the given number according to the various sets of the real numbers to which it belongs. (State all that apply.)

a. 0

b. 0.75

c. −25

d. $\sqrt{36}$

e. $\sqrt[3]{5}$

f. $\frac{2}{3}$

Solution

Step 1. Recall the various sets of numbers that make up the real numbers: natural numbers, whole numbers, integers, rational numbers, irrational numbers, and real numbers.

a. 0

Step 2. Categorize 0 according to its membership in the various sets.

0 is a whole number, an integer, a rational number, and a real number.

b. 0.75

Step 2. Categorize 0.75 according to its membership in the various sets.

0.75 is a rational number and a real number.

c. –25

Step 2. Categorize –25 according to its membership in the various sets.

–25 is an integer, a rational number, and a real number.

d. $\sqrt{36}$

Step 2. Categorize $\sqrt{36}$ according to its membership in the various sets.

$\sqrt{36} = 6$ is a natural number, a whole number, an integer, a rational number, and a real number.

e. $\sqrt[3]{5}$

Step 2. Categorize $\sqrt[3]{5}$ according to its membership in the various sets.

$\sqrt[3]{5}$ is an irrational number and a real number.

f. $\frac{2}{3}$

Step 2. Categorize $\frac{2}{3}$ according to its membership in the various sets.

$\frac{2}{3}$ is a rational number and a real number.

Problem Graph the real numbers $-4, -2.5, 0, \frac{3}{4}, \sqrt{3}, e$, and 3.6.

Solution

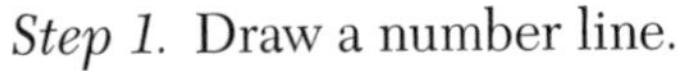

Step 1. Draw a number line.

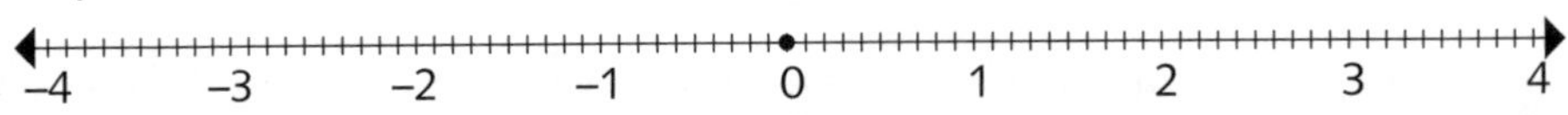

Step 2. Mark a large dot at each of the points corresponding to -4, -2.5, 0, $\frac{3}{4}$, $\sqrt{3}$, e, and 3.6. (Use $\sqrt{3} \approx 1.73$ and $e \approx 2.71$.)

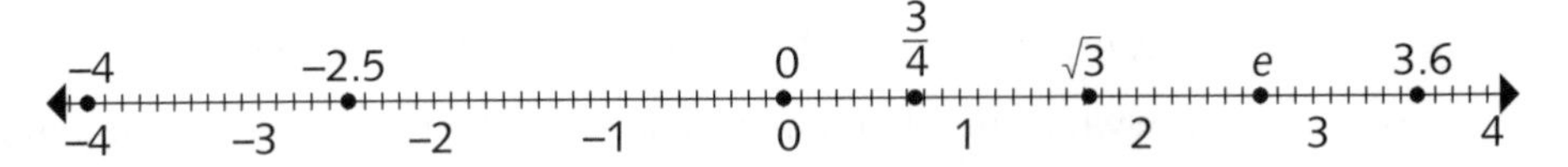

Properties of the Real Numbers

For much of algebra, you work with the set of real numbers along with the binary operations of *addition* and *multiplication*. A *binary operation* is one that you do on only two numbers at a time. Addition is indicated by the + sign. You can indicate multiplication a number of ways: For any two real numbers a and b, you can show a times b as $a \cdot b$, ab, $a(b)$, $(a)b$, or $(a)(b)$.

> Generally, in algebra, you do not use the times symbol × to indicate multiplication. This symbol is used when doing arithmetic.

The set of real numbers has the following 11 *field properties* for all real numbers a, b, and c under the operations of addition and multiplication.

1. **Closure Property of Addition.** $(a + b)$ is a real number. This property guarantees that the sum of any two real numbers is always a real number.

   Examples

   $(4 + 5)$ is a real number.

   $\left(\frac{1}{2} + \frac{3}{4}\right)$ is a real number.

   $(0.54 + 6.1)$ is a real number.

   $(\sqrt{2} + 1)$ is a real number.

2. **Closure Property of Multiplication.** $(a \cdot b)$ is a real number. This property guarantees that the product of any two real numbers is always a real number.

Examples

$(2 \cdot 7)$ is a real number.

$\left(\frac{1}{3} \cdot \frac{5}{8}\right)$ is a real number.

$[(2.5)(10.35)]$ is a real number.

$\left(\frac{1}{2} \cdot \sqrt{3}\right)$ is a real number.

3. **Commutative Property of Addition.** $a + b = b + a$. This property allows you to reverse the order of the numbers when you add, without changing the sum.

Examples

$4+5=5+4=9$

$\frac{1}{2}+\frac{3}{4}=\frac{3}{4}+\frac{1}{2}=\frac{5}{4}$

$0.54+6.1=6.1+0.54=6.64$

$\sqrt{2}+3\sqrt{2}=3\sqrt{2}+\sqrt{2}=4\sqrt{2}$

4. **Commutative Property of Multiplication.** $a \cdot b = b \cdot a$. This property allows you to reverse the order of the numbers when you multiply, without changing the product.

Examples

$2 \cdot 7=7 \cdot 2=14$

$\frac{1}{3} \cdot \frac{5}{8}=\frac{5}{8} \cdot \frac{1}{3}=\frac{5}{24}$

$(2.5)(10.35)=(10.35)(2.5)=25.875$

$\frac{1}{2} \cdot \sqrt{3}=\sqrt{3} \cdot \frac{1}{2}=\frac{\sqrt{3}}{2}$

5. **Associative Property of Addition.** $(a + b) + c = a + (b + c)$. This property says that when you have three numbers to add together, the final sum will be the same regardless of the way you group the numbers (two at a time) to perform the addition.

Example

Suppose you want to compute $6 + 3 + 7$. In the order given, you have two ways to group the numbers for addition:

$(6 + 3) + 7 = 9 + 7 = 16$ or $6 + (3 + 7) = 6 + 10 = 16$

Either way, 16 is the final sum.

6. **Associative Property of Multiplication.** $(ab)c = a(bc)$. This property says that when you have three numbers to multiply together, the final product will be the same regardless of the way you group the numbers (two at a time) to perform the multiplication.

The associative property is needed when you have to add or multiply more than two numbers because you can do addition or multiplication on only two numbers at a time. Thus, when you have three numbers, you must decide which two numbers you want to start with—the first two or the last two (assuming you keep the same order). Either way, your final answer is the same.

Example

Suppose you want to compute $7 \cdot 2 \cdot \frac{1}{2}$. In the order given, you have two ways to group the numbers for multiplication:

$$(7 \cdot 2) \cdot \frac{1}{2} = 14 \cdot \frac{1}{2} = 7 \text{ or } 7 \cdot (2 \cdot \frac{1}{2}) = 7 \cdot 1 = 7$$

Either way, 7 is the final product.

7. **Additive Identity Property.** There exists a real number 0, called the additive identity, such that $a + 0 = a$ and $0 + a = a$. This property guarantees that you have a real number, namely, 0, for which its sum with any real number is the number itself.

Examples

$$-8 + 0 = 0 + -8 = -8$$

$$\frac{5}{6} + 0 = 0 + \frac{5}{6} = \frac{5}{6}$$

8. **Multiplicative Identity Property.** There exists a real number 1, called the multiplicative identity, such that $a \cdot 1 = a$ and $1 \cdot a = a$. This property guarantees that you have a real number, namely, 1, for which its product with any real number is the number itself.

Examples

$$\sqrt{5} \cdot 1 = 1 \cdot \sqrt{5} = \sqrt{5}$$

$$-\frac{7}{8} \cdot 1 = 1 \cdot -\frac{7}{8} = -\frac{7}{8}$$

9. **Additive Inverse Property.** For every real number a, there is a real number called its additive inverse, denoted $-a$, such that $a + -a = 0$ and $-a + a = 0$. This property guarantees that every real number has an additive inverse (its opposite) that is a real number whose sum with the number is 0.

Examples

$6+-6=-6+6=0$

$7.43+-7.43=-7.43+7.43=0$

10. **Multiplicative Inverse Property.** For every *nonzero* real number a, there is a real number called its multiplicative inverse, denoted a^{-1} or $\frac{1}{a}$, such that $a \cdot a^{-1} = a \cdot \frac{1}{a} = 1$ and $a^{-1} \cdot a = \frac{1}{a} \cdot a = 1$. This property guarantees that every real number, *except zero*, has a multiplicative inverse (its reciprocal) whose product with the number is 1.

Notice that when you add the additive inverse to a number, you get the additive identity as an answer, and when you multiply a number by its multiplicative inverse, you get the multiplicative identity as an answer.

11. **Distributive Property.** $a(b + c) = a \cdot b + a \cdot c$ and $(b + c)\,a = b \cdot a + c \cdot a$. This property says that when you have a number times a sum (or a sum times a number), you can either add first and then multiply, or multiply first and then add. Either way, the final answer is the same.

Examples

$3(10+5)$ can be computed two ways:

add first to obtain $3(10+5)=3\cdot 15=45$ or

multiply first to obtain $3(10+5)=3\cdot 10+3\cdot 5=30+15=45$

Either way, the answer is 45.

$\left(\frac{1}{4}+\frac{3}{4}\right)8$ can be computed two ways:

add first to obtain $\left(\frac{1}{4}+\frac{3}{4}\right)8=1\cdot 8=8$ or

multiply first to obtain $\left(\frac{1}{4}+\frac{3}{4}\right)8=$

$\frac{1}{4}\cdot 8+\frac{3}{4}\cdot 8=2+6=8$

Either way, the answer is 8.

The distributive property is the only field property that involves both addition and multiplication at the same time. Another way to express the distributive property is to say that *multiplication distributes over addition.*

Problem State the field property that is illustrated in each of the following.

a. $0+1.25=1.25$

b. $(\pi+\sqrt{2}) \in$ real numbers

The symbol $\in$ is read "is an element of."

c. $\frac{3}{4} \cdot \frac{5}{6} = \frac{5}{6} \cdot \frac{3}{4}$

Solution

Step 1. Recall the 11 field properties: closure property of addition, closure property of multiplication, commutative property of addition, commutative property of multiplication, associative property of addition, associative property of multiplication, additive identity property, multiplicative identity property, additive inverse property, multiplicative inverse property, and distributive property.

a. $0 + 1.25 = 1.25$

Step 2. Identify the property illustrated.

Additive identity property

b. $(\pi + \sqrt{2}) \in$ real numbers

Step 2. Identify the property illustrated.

Closure property of addition

c. $\frac{3}{4} \cdot \frac{5}{6} = \frac{5}{6} \cdot \frac{3}{4}$

Step 2. Identify the property illustrated.

Commutative property of multiplication

Besides the field properties, you should keep in mind that the number 0 has the following unique characteristic.

12. **Zero Factor Property.** If a real number is multiplied by 0, the product is 0 (i.e., $a \cdot 0 = 0 \cdot a = 0$); and if the product of two numbers is 0, then at least one of the numbers is 0.

Examples

$$-9 \cdot 0 = 0 \cdot -9 = 0$$

$$\frac{15}{100} \cdot 0 = 0 \cdot \frac{15}{100} = 0$$

This property explains why 0 does not have a multiplicative inverse. There is no number that multiplies by 0 to give 1—because any number multiplied by 0 is 0.

Exercise 1

For 1–10, list all the sets in the real number system to which the given number belongs. (State all that apply.)

1. 10
2. $\sqrt{0.64}$
3. $\sqrt[3]{\dfrac{8}{125}}$
4. $-\pi$
5. -1000
6. $\sqrt{2}$
7. $-\sqrt{\dfrac{3}{4}}$
8. $-\sqrt{\dfrac{9}{4}}$
9. 1
10. $\sqrt[3]{0.001}$

For 11–20, state the property of the real numbers that is illustrated.

11. $\left(\dfrac{1}{4}\cdot 1500\right) \in$ real numbers
12. $\dfrac{2}{5}+\dfrac{3}{4}=\dfrac{3}{4}+\dfrac{2}{5}$
13. $43\cdot\dfrac{1}{43}=1$
14. $\left(1.3+\dfrac{1}{3}\right)$ is a real number
15. $43+(7+25)=(43+7)+25$
16. $60(10+3)=600+180=780$
17. $-\sqrt{41}+\sqrt{41}=0$
18. $-999\cdot 0=0$
19. $\left(0.6\cdot\dfrac{3}{4}\right)\dfrac{4}{3}=0.6\left(\dfrac{3}{4}\cdot\dfrac{4}{3}\right)$
20. $(90.75)(1)=90.75$

2

Computation with Real Numbers

This chapter presents the rules for computing with real numbers—often called signed numbers. Before proceeding with addition, subtraction, multiplication, and division of signed numbers, the discussion begins with comparing numbers and finding the absolute value of a number.

Comparing Numbers and Absolute Value

Comparing numbers uses the inequality symbols shown in Table 2.1.

Table 2.1 Inequality Symbols

INEQUALITY SYMBOL	*EXAMPLE*	*READ AS*
$<$	$2 < 7$	"2 is less than 7"
$>$	$7 > 2$	"7 is greater than 2"
$\leq$	$9 \leq 9$	"9 is less than or equal to 9"
$\geq$	$5 \geq 4$	"5 is greater than or equal to 4"
$\neq$	$2 \neq 7$	"2 is not equal to 7"

Graphing the numbers on a number line is helpful when you compare two numbers. The number that is farther to the right is the greater number. If the numbers coincide, they are equal; otherwise, they are unequal.

Problem Which is greater –7 or –2?

Solution

Step 1. Graph –7 and –2 on a number line.

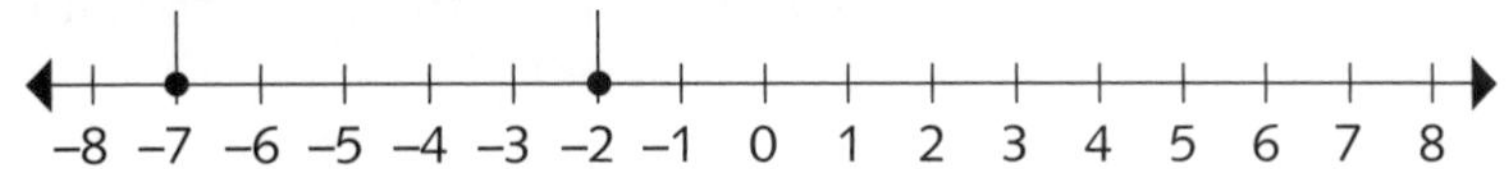

Step 2. Identify the number that is farther to the right as the greater number.

–2 is to the right of –7, so $-2 > -7$.

The concept of absolute value plays an important role in computations with signed numbers. The *absolute value* of a real number is its distance from 0 on the number line. For example, as shown in Figure 2.1, the absolute value of –8 is 8 because –8 is 8 units from 0.

Absolute value is a distance, so it is *never* negative.

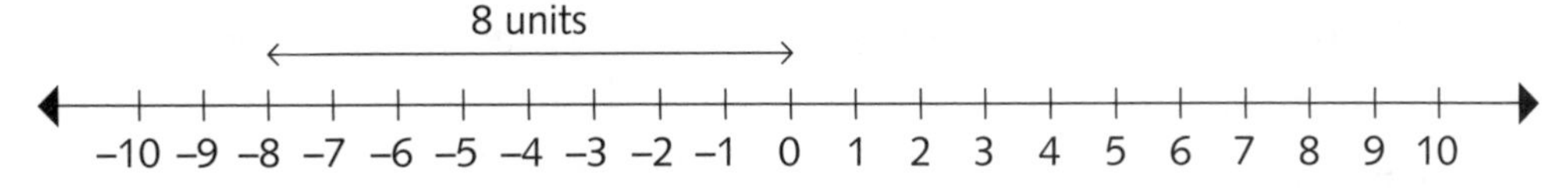

Figure 2.1 The absolute value of –8

You indicate the absolute value of a number by placing the number between a pair of vertical bars like this: $|-8|$ (read as "the absolute value of negative eight"). Thus, $|-8| = 8$.

Problem Find the indicated absolute value.

a. $|-30|$

b. $|0.4|$

c. $\left|-2\frac{1}{3}\right|$

Solution

a. $|-30|$

Step 1. Recalling that the absolute value of a real number is its distance from 0 on the number line, determine the absolute value.

$|-30| = 30$ because –30 is 30 units from 0 on the number line.

b. $|0.4|$

Step 1. Recalling that the absolute value of a real number is its distance from 0 on the number line, determine the absolute value.

$|0.4| = 0.4$ because 0.4 is 0.4 units from 0 on the number line.

c. $\left|-2\frac{1}{3}\right|$

Step 1. Recalling that the absolute value of a real number is its distance from 0 on the number line, determine the absolute value.

> As you likely noticed, the absolute value of a number is the value of the number with no sign attached. This strategy works for a number whose value you know, but do not use it when you don't know the value of the number.

$\left|-2\frac{1}{3}\right| = 2\frac{1}{3}$ because $-2\frac{1}{3}$ is $2\frac{1}{3}$ units from 0 on the number line.

Problem Which number has the greater absolute value?

a. $-35, 60$

b. $35, -60$

c. $-\frac{2}{9}, \frac{7}{9}$

d. $\frac{2}{9}, -\frac{7}{9}$

Solution

a. $-35, 60$

Step 1. Determine the absolute values.

$|-35| = 35, \ |60| = 60$

Step 2. Compare the absolute values.

60 has the greater absolute value because $60 > 35$.

b. $35, -60$

Step 1. Determine the absolute values.

$|35| = 35, \ |-60| = 60$

Step 2. Compare the absolute values.

-60 has the greater absolute value because $60 > 35$.

c. $-\frac{2}{9}, \frac{7}{9}$

Step 1. Determine the absolute values.

$$\left|-\frac{2}{9}\right| = \frac{2}{9}, \quad \left|\frac{7}{9}\right| = \frac{7}{9}$$

Step 2. Compare the absolute values.

$\frac{7}{9}$ has the greater absolute value because $\frac{7}{9} > \frac{2}{9}$.

d. $\frac{2}{9}, -\frac{7}{9}$

Step 1. Determine the absolute values.

$$\left|\frac{2}{9}\right| = \frac{2}{9}, \quad \left|-\frac{7}{9}\right| = \frac{7}{9}$$

Step 2. Compare the absolute values.

$-\frac{7}{9}$ has the greater absolute value because $\frac{7}{9} > \frac{2}{9}$.

Don't make the mistake of trying to compare the numbers without first finding the absolute values.

Addition and Subtraction of Signed Numbers

Real numbers are called signed numbers because these numbers may be positive, negative, or 0. From your knowledge of arithmetic, you already know how to do addition, subtraction, multiplication, and division with positive numbers and 0. To do these operations with all signed numbers, you simply use the absolute values of the numbers and follow these eight rules.

Addition of Signed Numbers

Rule 1. To add two numbers that have the same sign, add their absolute values and give the sum their common sign.

Rule 2. To add two numbers that have opposite signs, subtract the lesser absolute value from the greater absolute value and give the sum the sign of the number with the greater absolute value; if the two numbers have the same absolute value, their sum is 0.

Rule 3. The sum of 0 and any number is the number.

These rules might sound complicated, but practice will make them your own. One helpful hint is that when you need the absolute value of a number, just use the value of the number with no sign attached.

Problem Find the sum.

a. $-35+-60$

b. $35+-60$

c. $-35+60$

d. $-\frac{2}{9}+\frac{7}{9}$

e. $\frac{2}{9}+-\frac{7}{9}$

f. $-9.75+-8.12$

g. $-990.36+0$

Solution

a. $-35+-60$

Step 1. Determine which addition rule applies.

$-35+-60$

The signs are the same (both negative), so use Rule 1.

Step 2. Add the absolute values, 35 and 60.

$35+60=95$

Step 3. Give the sum a negative sign (the common sign).

$-35+-60=-95$

b. $35+-60$

Step 1. Determine which addition rule applies.

$35+-60$

The signs are opposites (one positive and one negative), so use Rule 2.

Step 2. Subtract 35 from 60 because $|-60| > |35|$.

$60 - 35 = 25$

Step 3. Make the sum negative because –60 has the greater absolute value.

$35 + -60 = -25$

c. $-35 + 60$

Step 1. Determine which addition rule applies.

$-35 + 60$

The signs are opposites (one negative and one positive), so use Rule 2.

Step 2. Subtract 35 from 60 because $|60| > |-35|$.

$60 - 35 = 25$

Step 3. Keep the sum positive because 60 has the greater absolute value.

$-35 + 60 = 25$

d. $-\frac{2}{9} + \frac{7}{9}$

Step 1. Determine which addition rule applies.

$-\frac{2}{9} + \frac{7}{9}$

The signs are opposites (one positive and one negative), so use Rule 2.

Step 2. Subtract $\frac{2}{9}$ from $\frac{7}{9}$ because $\left|\frac{7}{9}\right| > \left|-\frac{2}{9}\right|$.

$\frac{7}{9} - \frac{2}{9} = \frac{5}{9}$

Step 3. Keep the sum positive because $\frac{7}{9}$ has the greater absolute value.

$-\frac{2}{9} + \frac{7}{9} = \frac{5}{9}$

e. $\frac{2}{9}+-\frac{7}{9}$

Step 1. Determine which addition rule applies.

$\frac{2}{9}+-\frac{7}{9}$

The signs are opposites (one positive and one negative), so use Rule 2.

Step 2. Subtract $\frac{2}{9}$ from $\frac{7}{9}$ because $\left|-\frac{7}{9}\right|>\left|\frac{2}{9}\right|$.

$\frac{7}{9}-\frac{2}{9}=\frac{5}{9}$

Step 3. Make the sum negative because $-\frac{7}{9}$ has the greater absolute value.

$\frac{2}{9}+-\frac{7}{9}=-\frac{5}{9}$

f. $-9.75+-8.12$

Step 1. Determine which addition rule applies.

$-9.75+-8.12$

The signs are the same (both negative), so use Rule 1.

Step 2. Add the absolute values 9.75 and 8.12.

$9.75+8.12=17.87$

Step 3. Give the sum a negative sign (the common sign).

$-9.75+-8.12=-17.87$

g. $-990.36+0$

Step 1. Determine which addition rule applies.

$-990.36+0$

0 is added to a number, so the sum is the number (Rule 3).

$-990.36+0=-990.36$

You subtract signed numbers by changing the subtraction problem to an addition problem in a special way, so that you can apply the rules for addition of signed numbers. Here is the rule.

Subtraction of Signed Numbers

Rule 4. To subtract two numbers, keep the first number and add the opposite of the second number.

To apply this rule, think of the minus sign, –, as "add the opposite of." In other words, "subtracting a number" and "adding the opposite of the number" give the same answer.

Problem Change the subtraction problem to an addition problem.

a. $-35 - 60$

b. $35 - 60$

c. $60 - 35$

d. $-35 - (-60)$

e. $0 - 60$

f. $-60 - 0$

Solution

a. $-35 - 60$

Step 1. Keep –35.

-35

Step 2. Add the opposite of 60.

$= -35 + -60$

b. $35 - 60$

Step 1. Keep 35.

35

Step 2. Add the opposite of 60.

$= 35 + -60$

c. $60 - 35$

Step 1. Keep 60.

60

Step 2. Add the opposite of 35.

$= 60 + -35$

d. $-35 - (-60)$

Step 1. Keep -35.

-35

Step 2. Add the opposite of -60.

$= -35 + 60$

e. $0 - 60$

Step 1. Keep 0.

0

Step 2. Add the opposite of 60.

$= 0 + -60$

f. $-60 - 0$

Step 1. Keep -60.

-60

Step 2. Add the opposite of 0.

$= -60 + 0$

Remember 0 is its own opposite.

Problem Find the difference.

a. $-35 - 60$

b. $35 - 60$

c. $60 - 35$

d. $-35 - (-60)$

e. $0 - (-60)$

f. $-60 - 0$

A helpful mnemonic to remember how to subtract signed numbers is "Keep, change, change." You *keep* the first number, you *change* minus to plus, and you *change* the second number to its opposite.

Solution

a. $-35-60$

Step 1. Keep -35 and add the opposite of 60.

$-35-60$

$=-35+-60$

Step 2. The signs are the same (both negative), so use Rule 1 for addition.

$=-95$

Step 3. Review the main results.

$-35-60=-35+-60=-95$

Cultivate the habit of reviewing your main results. Doing so will help you catch careless mistakes.

b. $35-60$

Step 1. Keep 35 and add the opposite of 60.

$35-60$

$=35+-60$

Step 2. The signs are opposites (one positive and one negative), so use Rule 2 for addition.

$=-25$

Step 3. Review the main results.

$35-60=35+-60=-25$

c. $60-35$

Step 1. Keep 60 and add the opposite of 35.

$60-35$

$=60+-35$

Step 2. The signs are opposites (one positive and one negative), so use Rule 2 for addition.

$=25$

Step 3. Review the main results.

$60-35=60+-35=25$

d. $-35-(-60)$

Step 1. Keep -35 and add the opposite of -60.

$$-35-(-60)$$
$$=-35+60$$

Step 2. The signs are opposites (one positive and one negative), so use Rule 2 for addition.

$$=25$$

Step 3. Review the main results.

$$-35-(-60)=-35+60=25$$

e. $0-(-60)$

Step 1. Keep 0 and add the opposite of -60.

$$0-(-60)$$
$$=0+60$$

Step 2. 0 is added to a number, so the sum is the number (Rule 3 for addition).

$$=60$$

Step 3. Review the main results.

$$0-(-60)=0+60=60$$

f. $-60-0$

Step 1. Keep -60 and add the opposite of 0.

$$-60-0$$
$$=-60+0$$

Step 2. 0 is added to a number, so the sum is the number (Rule 3 for addition).

$$=-60$$

Step 3. Review the main results.

$$-60-0=-60+0=-60$$

Notice that subtraction is *not* commutative. That is, in general, for real numbers a and b, $a-b \neq b-a$.

Before going on, it is important that you distinguish the various uses of the short horizontal – symbol. Thus far, this symbol has three uses: (1) as part of a number to show that the number is negative, (2) as an indicator to find the opposite of the number that follows, and (3) as the minus symbol indicating subtraction.

Problem Given the statement $\underset{(1)}{\underset{\uparrow}{-}}(\underset{(2)}{\underset{\uparrow}{-}}35)\underset{(3)}{\underset{\uparrow}{-}}60 = 35 + \underset{(4)}{\underset{\uparrow}{-}}60$

a. Describe the use of the – symbols at (1), (2), (3), and (4).

b. Express the statement $-(-35) - 60 = 35 + -60$ in words.

Solution

a. Describe the use of the – symbols at (1), (2), (3), and (4).

Step 1. Interpret each – symbol.

The – symbol at (1) is an indicator to find the opposite of –35.

The – symbol at (2) is part of the number –35 that shows –35 is negative.

The – symbol at (3) is the minus symbol indicating subtraction.

The – symbol at (4) is part of the number –60 that shows –60 is negative.

Don't make the error of referring to negative numbers as "minus numbers."

The minus symbol always has a number to its immediate left.

There is never a number to the immediate left of a negative sign.

b. Express the statement $-(-35) - 60 = 35 + -60$ in words.

Step 1. Translate the statement into words.

$-(-35) - 60 = 35 + -60$ is read "the opposite of negative thirty-five minus sixty is thirty-five plus negative sixty."

Multiplication and Division of Signed Numbers

For multiplication of signed numbers, use the following three rules:

Multiplication of Signed Numbers

Rule 5. To multiply two numbers that have the same sign, multiply their absolute values and keep the product positive.

Rule 6. To multiply two numbers that have opposite signs, multiply their absolute values and make the product negative.

Rule 7. The product of 0 and any number is 0.

When you multiply two positive or two negative numbers, the product is *always* positive no matter what. Similarly, when you multiply two numbers that have opposite signs, the product is *always* negative—it doesn't matter which number has the greater absolute value.

Problem Find the product.

a. $(-3)(-40)$

b. $(3)(40)$

c. $(-3)(40)$

d. $(3)(-40)$

e. $(358)(0)$

Solution

a. $(-3)(-40)$

Step 1. Determine which multiplication rule applies.

$(-3)(-40)$

The signs are the same (both negative), so use Rule 5.

Step 2. Multiply the absolute values, 3 and 40.

$(3)(40)=120$

Step 3. Keep the product positive.

$(-3)(-40)=120$

b. $(3)(40)$

Step 1. Determine which multiplication rule applies.

$(3)(40)$

The signs are the same (both positive), so use Rule 5.

Step 2. Multiply the absolute values, 3 and 40.

$(3)(40)=120$

Step 3. Keep the product positive.

$(3)(40) = 120$

c. (–3)(40)

Step 1. Determine which multiplication rule applies.

$(-3)(40)$

The signs are opposites (one negative and one positive), so use Rule 6.

Step 2. Multiply the absolute values, 3 and 40.

$(3)(40) = 120$

Step 3. Make the product negative.

$(-3)(40) = -120$

d. (3)(–40)

Step 1. Determine which multiplication rule applies.

$(3)(-40)$

The signs are opposites (one positive and one negative), so use Rule 6.

Step 2. Multiply the absolute values, 3 and 40.

$(3)(40) = 120$

Step 3. Make the product negative.

$(3)(-40) = -120$

e. (358)(0)

Step 1. Determine which multiplication rule applies.

$(358)(0)$

0 is one of the factors, so use Rule 7.

Step 2. Find the product.

$(358)(0) = 0$

Rules 5, 6, and 7 tell you how to multiply two numbers, but often you will want to find the product of more than two numbers. To do this, multiply in pairs. You can keep track of the sign as you go along, or you simply can use the following guideline:

When 0 is one of the factors, the product is *always* 0; otherwise, products that have an even number of *negative* factors are positive, whereas those that have an odd number of *negative* factors are negative.

> Notice that if there is no zero factor, then the sign of the product is determined by how many *negative* factors you have.

Problem Find the product.

a. $(600)(-40)(-1000)(0)(-30)$

b. $(-3)(-10)(-5)(25)(-1)(-2)$

c. $(-2)(-4)(-10)(1)(-20)$

Solution

a. $(600)(-40)(-1000)(0)(-30)$

Step 1. 0 is one of the factors, so the product is 0.

$$(600)(-40)(-1000)(0)(-30) = 0$$

b. $(-3)(-10)(-5)(25)(-1)(-2)$

Step 1. Find the product ignoring the signs.

$$(3)(10)(5)(25)(1)(2) = 7500$$

Step 2. You have five negative factors, so make the product negative.

$$(-3)(-10)(-5)(25)(-1)(-2) = -7500$$

c. $(-2)(-4)(-10)(1)(-20)$

Step 1. Find the product ignoring the signs.

$$(2)(4)(10)(1)(20) = 1600$$

Step 2. You have four negative factors, so leave the product positive.

$$(-2)(-4)(-10)(1)(-20) = 1600$$

Division of Signed Numbers

Rule 8. To divide two numbers, divide their absolute values (being careful to make sure you don't divide by 0) and then follow the rules for multiplication of signed numbers.

If 0 is the dividend, the quotient is 0. For instance, $\frac{0}{5} = 0$. But if 0 is the divisor, the quotient is undefined. Thus, $\frac{5}{0} \neq 0$ and $\frac{5}{0} \neq 5$. $\frac{5}{0}$ has no answer because division by 0 is undefined!

Problem Find the quotient.

a. $\frac{-120}{-3}$

b. $\frac{-120}{3}$

c. $\frac{120}{-3}$

d. $\frac{-120}{0}$

e. $\frac{0}{30}$

In algebra, division is commonly indicated by the fraction bar.

Solution

a. $\frac{-120}{-3}$

Step 1. Divide 120 by 3.

$$\frac{120}{3} = 40$$

Step 2. The signs are the same (both negative), so keep the quotient positive.

$$\frac{-120}{-3} = 40$$

b. $\frac{-120}{3}$

Step 1. Divide 120 by 3.

$$\frac{120}{3} = 40$$

Step 2. The signs are opposites (one negative and one positive), so make the quotient negative.

$$\frac{-120}{3} = -40$$

c. $\frac{120}{-3}$

Step 1. Divide 120 by 3.

$$\frac{120}{3} = 40$$

Step 2. The signs are opposites (one positive and one negative), so make the quotient negative.

$$\frac{120}{-3} = -40$$

d. $\frac{-120}{0}$

Step 1. The divisor (denominator) is 0, so the quotient is undefined.

$$\frac{-120}{0} = \text{undefined}$$

e. $\frac{0}{30}$

Step 1. The dividend (numerator) is 0, so the quotient is 0.

$$\frac{0}{30} = 0$$

To be successful in algebra, you must memorize the rules for adding, subtracting, multiplying, and dividing signed numbers. Of course, when you do a computation, you don't have to write out all the steps. For instance, you can mentally ignore the signs to obtain the absolute values, do the necessary computation or computations, and then make sure your result has the correct sign.

Exercise 2

For 1–3, simplify.

1. $|-45|$
2. $|5.8|$
3. $\left|-5\frac{2}{3}\right|$

For 4 and 5, state in words.

4. $-9+-(-4)=-9+4$
5. $-9-(-4)=-9+4$

For 6–20, compute as indicated.

6. $-80+-40$
7. $0.7+-1.4$
8. $\left(-\frac{5}{6}\right)\left(\frac{2}{5}\right)$
9. $\frac{18}{-3}$
10. $(-100)(-8)$
11. $(400)\left(\frac{1}{2}\right)$
12. $\frac{-1\frac{1}{3}}{-\frac{1}{3}}$
13. $-450.95-(-65.83)$
14. $\frac{3}{11}-\left(-\frac{5}{11}\right)$
15. $\frac{0.8}{-0.01}$
16. $-458+0$
17. $\left(4\frac{1}{2}\right)\left(-3\frac{3}{5}\right)(0)(999)\left(-\frac{5}{17}\right)$
18. $\frac{0}{8.75}$
19. $\frac{700}{0}$
20. $(-3)(1)(-1)(-5)(-2)(2)(-10)$

3

Roots and Radicals

In this chapter, you learn about square roots, cube roots, and so on. Additionally, you learn about radicals and their relationship to roots. It is important in algebra that you have a facility for working with roots and radicals.

Squares, Square Roots, and Perfect Squares

You *square* a number by multiplying the number by itself. For instance, the square of 4 is $4 \cdot 4 = 16$. Also, the square of -4 is $-4 \cdot -4 = 16$. Thus, 16 is the result of squaring 4 or -4. The reverse of squaring is *finding the square root*. The two square roots of 16 are 4 and -4. You use the symbol $\sqrt{16}$ to represent the positive square root of 16. Thus, $\sqrt{16} = 4$. This number is the *principal square root* of 16. Thus, the principal square root of 16 is 4. Using the square root notation, you indicate the negative square root of 16 as $-\sqrt{16}$. Thus, $-\sqrt{16} = -4$.

$\sqrt{-16} \neq -4$; $\sqrt{-16}$ is not a real number because no real number multiplies by itself to give -16.

Every positive number has two square roots that are equal in absolute value, but opposite in sign. The positive square root is called the *principal square root* of the number. The number 0 has only one square root, namely, 0. The principal square root of 0 is 0. In general, if x is a real number such that $x \cdot x = s$, then $\sqrt{s} = |x|$ (the absolute value of x).

The $\sqrt{\ }$ symbol *always* gives *one* number as the answer and that number is nonnegative: positive or 0.

A number that is an exact square of another number is a *perfect square.* For instance, the integers 4, 9, 16, and 25 are perfect squares. Here is a helpful list of principal square roots of some perfect squares.

$\sqrt{0} = 0,\ \sqrt{1} = 1,\ \sqrt{4} = 2,\ \sqrt{9} = 3,\ \sqrt{16} = 4,$

$\sqrt{25} = 5,\ \sqrt{36} = 6,\ \sqrt{49} = 7,$

$\sqrt{64} = 8,\ \sqrt{81} = 9,\ \sqrt{100} = 10,\ \sqrt{121} = 11,$

$\sqrt{144} = 12,\ \sqrt{169} = 13,\ \sqrt{196} = 14,$

$\sqrt{225} = 15,\ \sqrt{256} = 16,\ \sqrt{289} = 17,$

$\sqrt{400} = 20,\ \sqrt{625} = 25$

> Working with square roots will be much easier for you if you memorize the list of square roots. Make flashcards to help you do this.

Also, fractions and decimals can be perfect squares. For instance, $\frac{9}{25}$ is a perfect square because $\frac{9}{25}$ equals $\frac{3}{5} \cdot \frac{3}{5}$, and 0.36 is a perfect square because 0.36 equals (0.6)(0.6). If a number is not a perfect square, you can indicate its square roots by using the square root symbol. For instance, the two square roots of 15 are $\sqrt{15}$ and $-\sqrt{15}$.

Problem Find the two square roots of the given number.

a. 25

b. $\frac{4}{9}$

c. 0.49

d. 11

Solution

a. 25

Step 1. Find the principal square root of 25.

$5 \cdot 5 = 25$, so 5 is the principal square root of 25.

Step 2. Write the two square roots of 25.

5 and –5 are the two square roots of 25.

b. $\frac{4}{9}$

Step 1. Find the principal square root of $\frac{4}{9}$.

$\frac{2}{3} \cdot \frac{2}{3} = \frac{4}{9}$, so $\frac{2}{3}$ is the principal square root of $\frac{4}{9}$.

Step 2. Write the two square roots of $\frac{4}{9}$.

$\frac{2}{3}$ and $-\frac{2}{3}$ are the two square roots of $\frac{4}{9}$.

c. 0.49

Step 1. Find the principal square root of 0.49.

$(0.7)(0.7) = 0.49$, so 0.7 is the principal square root of 0.49.

Step 2. Write the two square roots of 0.49.

0.7 and –0.7 are the two square roots of 0.49.

d. 11

Step 1. Find the principal square root of 11.

$\sqrt{11}$ is the principal square root of 11.

Because 11 is not a perfect square, you indicate the square root.

Step 2. Write the two square roots of 11.

$\sqrt{11}$ and $-\sqrt{11}$ are the two square roots of 11.

Problem Find the indicated root.

a. $\sqrt{81}$

b. $\sqrt{100}$

c. $\sqrt{\frac{4}{25}}$

d. $\sqrt{30}$

e. $\sqrt{9+16}$

f. $\sqrt{-2 \cdot -2}$

g. $\sqrt{b \cdot b}$

Solution

a. $\sqrt{81}$

Step 1. Find the principal square root of 81.

$$\sqrt{81} = 9$$

> $\sqrt{81} \neq \pm 9$. The square root symbol *always* gives just *one* nonnegative number as the answer! If you want ± 9, then do this: $\pm\sqrt{81} = \pm 9$.

b. $\sqrt{100}$

Step 1. Find the principal square root of 100.

$$\sqrt{100} = 10$$

> $\sqrt{100} \neq 50$. You do not divide by 2 to get a square root.

c. $\sqrt{\frac{4}{25}}$

Step 1. Find the principal square root of $\frac{4}{25}$.

$$\sqrt{\frac{4}{25}} = \frac{2}{5}$$

d. $\sqrt{30}$

Step 1. Find the principal square root of 30.

Because 30 is not a perfect square, $\sqrt{30}$ indicates the principal square root of 30.

e. $\sqrt{9+16}$

Step 1. Add 9 and 16 because you want the principal square root of the quantity $9+16$. (See Chapter 5 for a discussion of $\sqrt{}$ as a grouping symbol.)

$$\sqrt{9+16} = \sqrt{25}$$

Step 2. Find the principal square root of 25.

$$\sqrt{9+16} = \sqrt{25} = 5$$

> $\sqrt{9+16} \neq \sqrt{9} + \sqrt{16}$. $\sqrt{9+16} = \sqrt{25} = 5$, but $\sqrt{9} + \sqrt{16} = 3 + 4 = 7$.

f. $\sqrt{-2 \cdot -2}$

Step 1. Find the principal square root of $-2 \cdot -2$.

$$\sqrt{-2 \cdot -2} = |-2| = 2$$

$\sqrt{-2 \cdot -2} \neq -2$. The $\sqrt{\ }$ symbol *never* gives a negative number as an answer.

g. $\sqrt{b \cdot b}$

Step 1. Find the principal square root of $b \cdot b$.

$$\sqrt{b \cdot b} = |b|$$

$\sqrt{b \cdot b} \neq b$ if b is negative and $|b| \neq b$ if b is negative. Because you don't know the value of the number b, you must keep the absolute value bars.

Cube Roots and *n*th Roots

A number x such that $x \cdot x \cdot x = c$ is a *cube root* of c. Finding the cube root of a number is the reverse of cubing a number. Every real number has exactly *one* real cube root, called its *principal cube root*. For example, because $-4 \cdot -4 \cdot -4 = -64$, -4 is the principal cube root of -64. You use $\sqrt[3]{-64}$ to indicate the principal cube root of -64. Thus, $\sqrt[3]{-64} = -4$. Similarly, $\sqrt[3]{64} = 4$. As you can see, the principal cube root of a negative number is negative, and the principal cube root of a positive number is positive. In general, if x is a real number such that $x \cdot x \cdot x = c$, then $\sqrt[3]{c} = x$. Here is a list of principal cube roots of some *perfect cubes* that are useful to know.

$$\sqrt[3]{0} = 0,\ \sqrt[3]{1} = 1,\ \sqrt[3]{8} = 2,\ \sqrt[3]{27} = 3,$$
$$\sqrt[3]{64} = 4,\ \sqrt[3]{125} = 5,\ \sqrt[3]{1000} = 10$$

You will find it worth your while to memorize the list of cube roots.

If a number is not a perfect cube, you indicate its principal cube root by using the cube root symbol. For instance, the cube root of -18 is $\sqrt[3]{-18}$.

Problem Find the indicated root.

a. $\sqrt[3]{-27}$

b. $\sqrt[3]{\dfrac{8}{125}}$

c. $\sqrt[3]{0.008}$

d. $\sqrt[3]{-1}$

e. $\sqrt[3]{-7\cdot-7\cdot-7}$

f. $\sqrt[3]{b\cdot b\cdot b}$

Solution

a. $\sqrt[3]{-27}$

Step 1. Find the principal cube root of -27.

$$-3\cdot-3\cdot-3=-27,\text{ so }\sqrt[3]{-27}=-3.$$

$\sqrt[3]{-27}\neq-9$. You do not divide by 3 to get a cube root.

b. $\sqrt[3]{\dfrac{8}{125}}$

Step 1. Find the principal cube root of $\dfrac{8}{125}$.

$$\frac{2}{5}\cdot\frac{2}{5}\cdot\frac{2}{5}=\frac{8}{125},\text{ so }\sqrt[3]{\frac{8}{125}}=\frac{2}{5}.$$

c. $\sqrt[3]{0.008}$

Step 1. Find the principal cube root of 0.008.

$$(0.2)(0.2)(0.2)=0.008,\text{ so }\sqrt[3]{0.008}=0.2.$$

d. $\sqrt[3]{-1}$

Step 1. Find the principal cube root of -1.

$$-1\cdot-1\cdot-1=-1,\text{ so }\sqrt[3]{-1}=-1.$$

e. $\sqrt[3]{-7\cdot-7\cdot-7}$

Step 1. Find the principal cube root of $-7\cdot-7\cdot-7$.

$$\sqrt[3]{-7\cdot-7\cdot-7}=-7$$

f. $\sqrt[3]{b\cdot b\cdot b}$

Step 1. Find the principal cube root of $b\cdot b\cdot b$.

$$\sqrt[3]{b\cdot b\cdot b}=b$$

In general, if $\underbrace{x \cdot x \cdot x \cdot \cdots \cdot x}_{n \text{ factors of } x} = a$, where n is a natural number, x is called an *nth root* of a. The *principal nth root* of a is denoted $\sqrt[n]{a}$. The expression $\sqrt[n]{a}$ is called a *radical*, a is called the *radicand*, n is called the *index* and indicates which root is desired. If no index is written, it is understood to be 2 and the radical expression indicates the principal square root of the radicand. As a rule, a *positive* real number has exactly *one* real positive nth root whether n is even or odd, and *every* real number has exactly one real nth root when n is odd. Negative numbers do not have real nth roots when n is even. Finally, the nth root of 0 is 0 whether n is even or odd: $\sqrt[n]{0} = 0$ (always).

Problem Find the indicated root, if possible.

a. $\sqrt[4]{81}$

b. $\sqrt[5]{-\frac{1}{32}}$

c. $\sqrt[3]{0.125}$

d. $\sqrt[6]{-1}$

e. $\sqrt[7]{-1}$

f. $\sqrt[50]{0}$

Solution

a. $\sqrt[4]{81}$

Step 1. Find the principal fourth root of 81.

$$3 \cdot 3 \cdot 3 \cdot 3 = 81, \text{ so } \sqrt[4]{81} = 3.$$

b. $\sqrt[5]{-\frac{1}{32}}$

Step 1. Find the principal fifth root of $-\frac{1}{32}$.

$$-\frac{1}{2} \cdot -\frac{1}{2} \cdot -\frac{1}{2} \cdot -\frac{1}{2} \cdot -\frac{1}{2} = -\frac{1}{32}, \text{ so } \sqrt[5]{-\frac{1}{32}} = -\frac{1}{2}.$$

c. $\sqrt[3]{0.125}$

Step 1. Find the principal cube root of 0.125.

$$(0.5)(0.5)(0.5) = 0.125, \text{ so } \sqrt[3]{0.125} = 0.5.$$

d. $\sqrt[6]{-1}$

Step 1. -1 is negative and 6 is even, so $\sqrt[6]{-1}$ is not a real number.

$\sqrt[6]{-1}$ is not defined for real numbers.

$\sqrt[6]{-1} \neq -1. -1 \cdot -1 \cdot -1 \cdot -1 \cdot -1 \cdot -1 = 1$, not -1.

e. $\sqrt[7]{-1}$

Step 1. Find the principal seventh root of -1.

$$-1 \cdot -1 \cdot -1 \cdot -1 \cdot -1 \cdot -1 \cdot -1 = -1, \text{ so } \sqrt[7]{-1} = -1.$$

f. $\sqrt[50]{0}$

Step 1. Find the principal 50th root of 0.

The nth root of 0 is 0, so $\sqrt[50]{0} = 0$.

Simplifying Radicals

Sometimes in algebra you have to *simplify radicals*—most frequently, square root radicals. A square root radical is in simplest form when it has (a) no factors that are perfect squares and (b) no fractions. You use the following property of square root radicals to accomplish the simplifying.

If a and b are nonnegative numbers,
$\sqrt{a \cdot b} = \sqrt{a} \cdot \sqrt{b}$

Problem Simplify.

a. $\sqrt{48}$

b. $\sqrt{360}$

c. $\sqrt{\frac{3}{4}}$

d. $\sqrt{\frac{1}{2}}$

Solution

a. $\sqrt{48}$

Step 1. Express $\sqrt{48}$ as a product of two numbers, one of which is the largest perfect square.

$\sqrt{48}$

$= \sqrt{16 \cdot 3}$

Step 2. Replace $\sqrt{16 \cdot 3}$ with the product of the square roots of 16 and 3.

$= \sqrt{16} \cdot \sqrt{3}$

Step 3. Find $\sqrt{16}$ and put the answer in front of $\sqrt{3}$ as a coefficient. (See Chapter 6 for a discussion of the term *coefficient*.)

$= 4\sqrt{3}$

Step 4. Review the main results.

$\sqrt{48} = \sqrt{16 \cdot 3} = \sqrt{16} \cdot \sqrt{3} = 4\sqrt{3}$

b. $\sqrt{360}$

Step 1. Express $\sqrt{360}$ as a product of two numbers, one of which is the largest perfect square.

$\sqrt{360}$

$= \sqrt{36 \cdot 10}$

Step 2. Replace $\sqrt{36 \cdot 10}$ with the product of the square roots of 36 and 10.

$= \sqrt{36} \cdot \sqrt{10}$

Step 3. Find $\sqrt{36}$ and put the answer in front of $\sqrt{10}$ as a coefficient.

$$= 6\sqrt{10}$$

Step 4. Review the main results.

$$\sqrt{360} = \sqrt{36 \cdot 10} = \sqrt{36} \cdot \sqrt{10} = 6\sqrt{10}$$

c. $\sqrt{\dfrac{3}{4}}$

Step 1. Express $\sqrt{\dfrac{3}{4}}$ as a product of two numbers, one of which is the largest perfect square.

$$\sqrt{\frac{3}{4}}$$

$$= \sqrt{\frac{1}{4} \cdot 3}$$

Step 2. Replace $\sqrt{\dfrac{1}{4} \cdot 3}$ with the product of the square roots of $\dfrac{1}{4}$ and 3.

$$= \sqrt{\frac{1}{4}} \cdot \sqrt{3}$$

Step 3. Find $\sqrt{\dfrac{1}{4}}$ and put the answer in front of $\sqrt{3}$ as a coefficient.

$$= \frac{1}{2}\sqrt{3}$$

Step 4. Review the main results.

$$\sqrt{\frac{3}{4}} = \sqrt{\frac{1}{4} \cdot 3} = \sqrt{\frac{1}{4}} \cdot \sqrt{3} = \frac{1}{2}\sqrt{3}$$

d. $\sqrt{\frac{1}{2}}$

Step 1. Multiply the numerator and the denominator of $\frac{1}{2}$ by the least number that will make the denominator a perfect square.

$$\sqrt{\frac{1}{2}}$$

$$= \sqrt{\frac{1 \cdot 2}{2 \cdot 2}} = \sqrt{\frac{2}{4}}$$

Step 2. Express $\sqrt{\frac{2}{4}}$ as a product of two numbers, one of which is the largest perfect square.

$$= \sqrt{\frac{1}{4} \cdot 2}$$

Step 3. Replace $\sqrt{\frac{1}{4} \cdot 2}$ with the product of the square roots of $\frac{1}{4}$ and 2.

$$= \sqrt{\frac{1}{4}} \cdot \sqrt{2}$$

Step 4. Find $\sqrt{\frac{1}{4}}$ and put the answer in front of $\sqrt{2}$ as a coefficient.

$$= \frac{1}{2}\sqrt{2}$$

Step 5. Review the main results.

$$\sqrt{\frac{1}{2}} = \sqrt{\frac{1 \cdot 2}{2 \cdot 2}} = \sqrt{\frac{2}{4}} = \sqrt{\frac{1}{4} \cdot 2} = \sqrt{\frac{1}{4}} \cdot \sqrt{2} = \frac{1}{2}\sqrt{2}$$

Exercise 3

For 1–4, find the two square roots of the given number.

1. 144
2. $\frac{25}{49}$
3. 0.64
4. 400

For 5–18, find the indicated root, if possible.

5. $\sqrt{16}$
6. $\sqrt{-9}$
7. $\sqrt{\frac{16}{25}}$
8. $\sqrt{25+144}$
9. $\sqrt{-5 \cdot -5}$
10. $\sqrt{z \cdot z}$
11. $\sqrt[3]{-125}$
12. $\sqrt[3]{\frac{64}{125}}$
13. $\sqrt[3]{0.027}$
14. $\sqrt[3]{y \cdot y \cdot y}$
15. $\sqrt[4]{625}$
16. $\sqrt[5]{-\frac{32}{243}}$
17. $\sqrt[6]{-64}$
18. $\sqrt[7]{0}$

For 19 and 20, simplify.

19. $\sqrt{72}$
20. $\sqrt{\frac{2}{3}}$

Exponentiation

This chapter presents a detailed discussion of exponents. Working efficiently and accurately with exponents will serve you well in algebra.

Exponents

An *exponent* is a small raised number written to the upper right of a quantity, which is called the *base* for the exponent. For example, consider the product $3 \cdot 3 \cdot 3 \cdot 3 \cdot 3$, in which the same number is repeated as a factor multiple times. The shortened notation for $3 \cdot 3 \cdot 3 \cdot 3 \cdot 3$ is 3^5. This representation of the product is an *exponential expression*. The number 3 is the *base*, and the small 5 to the upper right of 3 is the *exponent*. Most commonly, the exponential expression 3^5 is read as "three to the fifth." Other ways you might read 3^5 are "three to the fifth power" or "three raised to the fifth power." In general, x^a is "x to the ath," "x to the ath power," or "x raised to the ath power."

Exponentiation is the act of evaluating an exponential expression, x^a.

> *Exponentiation* is a big word, but it just means that you do to the base what the exponent tells you to do to it.

The result you get is the ath *power* of the base. For instance, to evaluate 3^5, which has the natural number 5 as an exponent, perform the multiplication as shown here (see Figure 4.1).

Step 1. Write 3^5 in product form.

$$3^5 = 3 \cdot 3 \cdot 3 \cdot 3 \cdot 3$$

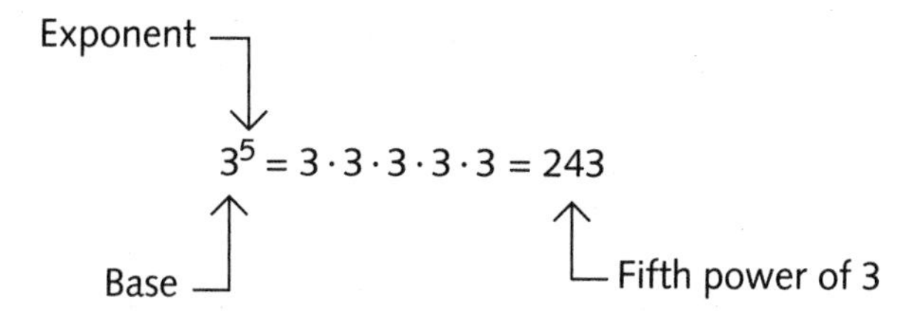

Figure 4.1 Parts of an exponential form

Step 2. Do the multiplication.

$$3^5 = 3 \cdot 3 \cdot 3 \cdot 3 \cdot 3 = 243 \text{ (the fifth power of 3)}$$

The following discussion tells you about the different types of exponents and what they tell you to do to the base.

Natural Number Exponents

You likely are most familiar with natural number exponents.

Natural Number Exponents

If x is a real number and n is a natural number, then $x^n = \underbrace{x \cdot x \cdot x \cdot \cdots \cdot x}_{n \text{ factors of } x}$.

For instance, 5^4 has a natural number exponent, namely, 4. The exponent 4 tells you how many times to use the base 5 as a factor. When you do the exponentiation, the product is the fourth power of 5 as shown in Figure 4.2.

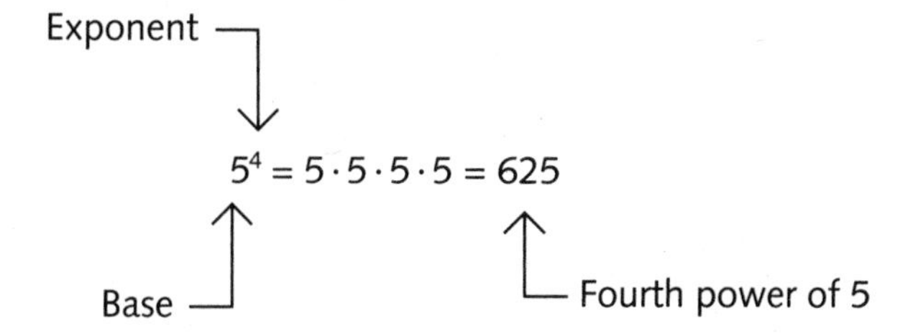

Figure 4.2 Fourth power of 5

For the first power of a number, for instance, 5^1, you usually omit the exponent and simply write 5. The second power of a number is the *square* of the number; read 5^2 as "five squared." The third power of a number is the *cube* of the number; read 5^3 as "five cubed." Beyond the third power, read 5^4 as "five to the fourth," read 5^5 as "five to the fifth," read 5^6 as "five to the sixth," and so on.

> Don't multiply the base by the exponent! $5^2 \neq 5 \cdot 2 = 10$, $5^3 \neq 5 \cdot 3 = 15$, $5^4 \neq 5 \cdot 4 = 20$, etc. $x^n \neq x \cdot n$.

Problem Write the indicated product as an exponential expression.

a. $2 \cdot 2 \cdot 2 \cdot 2 \cdot 2 \cdot 2 \cdot 2$

b. $-3 \cdot -3 \cdot -3 \cdot -3 \cdot -3 \cdot -3$

Solution

a. $2 \cdot 2 \cdot 2 \cdot 2 \cdot 2 \cdot 2 \cdot 2$

Step 1. Count how many times 2 is a factor.

$$\underbrace{2 \cdot 2 \cdot 2 \cdot 2 \cdot 2 \cdot 2 \cdot 2}_{\text{Seven factors of 2}}$$

Step 2. Write the indicated product as an exponential expression with 2 as the base and 7 as the exponent.

$$2 \cdot 2 \cdot 2 \cdot 2 \cdot 2 \cdot 2 \cdot 2 = 2^7$$

b. $-3 \cdot -3 \cdot -3 \cdot -3 \cdot -3 \cdot -3$

Step 1. Count how many times -3 is a factor.

$$\underbrace{-3 \cdot -3 \cdot -3 \cdot -3 \cdot -3 \cdot -3}_{\text{Six factors of } -3}$$

Step 2. Write the indicated product as an exponential expression with -3 as the base and 6 as the exponent.

$$-3 \cdot -3 \cdot -3 \cdot -3 \cdot -3 \cdot -3 = (-3)^6$$

In the above problem, you must enclose the -3 in parentheses to show that -3 is the number that is used as a factor six times. Only the 3 will be raised to the power unless parentheses are used to indicate otherwise.

> $(-3)^6 \neq -3^6$. $(-3)^6 = 729$, but $-3^6 = -729$.

Problem Evaluate.

a. 2^5

b. $(-2)^5$

c. $(0.6)^2$

d. $\left(\frac{3}{4}\right)^3$

e. 0^{100}

f. 1^3

g. $(1+1)^3$

Solution

a. 2^5

Step 1. Write 2^5 in product form.

$$2^5 = 2\cdot 2\cdot 2\cdot 2\cdot 2$$

Step 2. Do the multiplication.

$$2^5 = 2\cdot 2\cdot 2\cdot 2\cdot 2 = 32$$

b. $(-2)^5$

Step 1. Write $(-2)^5$ in product form.

$$(-2)^5 = -2\cdot -2\cdot -2\cdot -2\cdot -2$$

Step 2. Do the multiplication.

$$(-2)^5 = -2\cdot -2\cdot -2\cdot -2\cdot -2 = -32$$

c. $(0.6)^2$

Step 1. Write $(0.6)^2$ in product form.

$$(0.6)^2 = (0.6)(0.6)$$

Step 2. Do the multiplication.

$$(0.6)^2 = (0.6)(0.6) = 0.36$$

d. $\left(\frac{3}{4}\right)^3$

Step 1. Write $\left(\frac{3}{4}\right)^3$ in product form.

$$\left(\frac{3}{4}\right)^3 = \frac{3}{4}\cdot\frac{3}{4}\cdot\frac{3}{4}$$

Step 2. Do the multiplication.

$$\left(\frac{3}{4}\right)^3 = \frac{3}{4}\cdot\frac{3}{4}\cdot\frac{3}{4} = \frac{27}{64}$$

e. 0^{100}

Step 1. Because 0^{100} has 0 as a factor 100 times, the product is 0.

$$0^{100} = 0$$

f. 1^3

Step 1. Write 1^3 in product form.

$$1^3 = 1\cdot 1\cdot 1$$

Step 2. Do the multiplication.

$$1^3 = 1\cdot 1\cdot 1 = 1$$

g. $(1+1)^3$

Step 1. Add 1 and 1 because you want to cube the quantity $1+1$. (See Chapter 5 for a discussion of parentheses as a grouping symbol.)

$$(1+1)^3 = 2^3$$

Step 2. Write 2^3 in product form.

$$2^3 = 2\cdot 2\cdot 2$$

Step 3. Do the multiplication.

$$2^3 = 2\cdot 2\cdot 2 = 8$$

$(1+1)^3 \neq 1^3 + 1^3$. $(1+1)^3 = 2^3 = 8$, but $1^3 + 1^3 = 1 + 1 = 2$.

Zero and Negative Integer Exponents

Zero Exponent

If x is a nonzero real number, then $x^0 = 1$.

0^0 is undefined; it has no meaning. $0^0 \neq 0$.

A zero exponent on a nonzero number tells you to put 1 as the answer when you evaluate.

$x^0 \neq 0$; $x^0 = 1$, provided $x \neq 0$.

Problem Evaluate.

a. $(-2)^0$

b. $(0.6)^0$

c. $\left(\frac{3}{4}\right)^0$

d. π^0

e. 1^0

Solution

a. $(-2)^0$

Step 1. The exponent is 0, so the answer is 1.

$$(-2)^0 = 1$$

b. $(0.6)^0$

Step 1. The exponent is 0, so the answer is 1.

$$(0.6)^0 = 1$$

c. $\left(\frac{3}{4}\right)^0$

Step 1. The exponent is 0, so the answer is 1.

$$\left(\frac{3}{4}\right)^0 = 1$$

d. π^0

Step 1. The exponent is 0, so the answer is 1.

$$\pi^0 = 1$$

e. 1^0

Step 1. The exponent is 0, so the answer is 1.

$$1^0 = 1$$

Negative Integer Exponents

If x is a nonzero real number and n is a natural number, then $x^{-n} = \frac{1}{x^n}$.

A negative integer exponent on a nonzero number tells you to obtain the *reciprocal of the corresponding exponential expression that has a positive exponent.*

$x^{-n} \neq -\frac{1}{x^n}$; $x^{-n} \neq -x^n$. A negative exponent does not make a power negative.

Problem Evaluate.

a. 2^{-5}

b. $(-2)^{-5}$

c. $(0.6)^{-2}$

d. $\left(\frac{3}{4}\right)^{-3}$

Solution

a. 2^{-5}

Step 1. Write the reciprocal of the corresponding positive exponent version of 2^{-5}.

$$2^{-5} = \frac{1}{2^5}$$

Step 2. Evaluate 2^5.

$$2^{-5} = \frac{1}{2^5} = \frac{1}{2 \cdot 2 \cdot 2 \cdot 2 \cdot 2} = \frac{1}{32}$$

As you can see, the negative exponent did not make the answer negative; $2^{-5} \neq -\frac{1}{32}$.

b. $(-2)^{-5}$

Step 1. Write the reciprocal of the corresponding positive exponent version of $(-2)^{-5}$.

$$(-2)^{-5} = \frac{1}{(-2)^5}$$

When you evaluate $(-2)^{-5}$, the answer is negative because $(-2)^5$ is negative. The negative exponent is not the reason $(-2)^{-5}$ is negative.

Step 2. Evaluate $(-2)^5$.

$$(-2)^{-5} = \frac{1}{(-2)^5} = \frac{1}{-2 \cdot -2 \cdot -2 \cdot -2 \cdot -2} = \frac{1}{-32} = -\frac{1}{32}$$

c. $(0.6)^{-2}$

Step 1. Write the reciprocal of the corresponding positive exponent version of $(0.6)^{-2}$.

$$(0.6)^{-2} = \frac{1}{(0.6)^2}$$

Step 2. Evaluate $(0.6)^2$.

$$(0.6)^{-2} = \frac{1}{(0.6)^2} = \frac{1}{(0.6)(0.6)} = \frac{1}{0.36}$$

d. $\left(\frac{3}{4}\right)^{-3}$

Step 1. Write the reciprocal of the corresponding positive exponent version of $\left(\frac{3}{4}\right)^{-3}$.

$$\left(\frac{3}{4}\right)^{-3} = \frac{1}{(3/4)^3}$$

Step 2. Evaluate $\left(\frac{3}{4}\right)^{3}$ and simplify.

$$\left(\frac{3}{4}\right)^{-3} = \frac{1}{(3/4)^3} = \frac{1}{27/64} = \frac{64}{27}$$

Notice that because $x^{-n} = \frac{1}{x^n}$, the expression $\frac{1}{x^{-n}}$ can be simplified as follows:
$\frac{1}{x^{-n}} = \frac{1}{1/x^n} = \frac{x^n}{1} = x^n$; thus, $\frac{1}{x^{-n}} = x^n$. Apply this rule in the following problem.

Problem Simplify.

a. $\frac{1}{2^{-5}}$

b. $\frac{1}{(-2)^{-5}}$

c. $\frac{1}{(0.6)^{-2}}$

Solution

a. $\frac{1}{2^{-5}}$

Step 1. Apply $\frac{1}{x^{-n}} = x^n$.

$$\frac{1}{2^{-5}} = 2^5$$

$\frac{1}{2^{-5}} \neq \frac{1}{2}^5$. Keep the same base for the corresponding positive exponent version.

Step 2. Evaluate 2^5.

$$\frac{1}{2^{-5}} = 2^5 = 32$$

b. $\frac{1}{(-2)^{-5}}$

Step 1. Apply $\frac{1}{x^{-n}} = x^n$.

$$\frac{1}{(-2)^{-5}} = (-2)^5$$

Step 2. Evaluate $(-2)^5$.

$$\frac{1}{(-2)^{-5}} = (-2)^5 = -32$$

c. $\frac{1}{(0.6)^{-2}}$

Step 1. Apply $\frac{1}{x^{-n}} = x^n$.

$$\frac{1}{(0.6)^{-2}} = (0.6)^2$$

Step 2. Evaluate $(0.6)^2$.

$$\frac{1}{(0.6)^{-2}} = (0.6)^2 = 0.36$$

Unit Fraction and Rational Exponents

Unit Fraction Exponents

If x is a real number and n is a natural number, then $x^{1/n} = \sqrt[n]{x}$, provided that, when n is even, $x \geq 0$.

A unit fraction exponent on a number tells you to find the principal nth root of the number.

Problem Evaluate, if possible.

a. $(-27)^{1/3}$

b. $(0.25)^{1/2}$

c. $(-16)^{1/4}$

d. $\left(-\frac{32}{243}\right)^{1/5}$

Solution

a. $(-27)^{1/3}$

Step 1. Apply $x^{1/n} = \sqrt[n]{x}$.

$$(-27)^{1/3} = \sqrt[3]{-27}$$

Step 2. Find the principal cube root of -27.

$$(-27)^{1/3} = \sqrt[3]{-27} = -3$$

b. $(0.25)^{1/2}$

Step 1. Apply $x^{1/n} = \sqrt[n]{x}$.

$$(0.25)^{1/2} = \sqrt{0.25}$$

Step 2. Find the principal square root of 0.25.

$$(0.25)^{1/2} = \sqrt{0.25} = 0.5$$

c. $(-16)^{1/4}$

Step 1. Apply $x^{1/n} = \sqrt[n]{x}$.

$$(-16)^{1/4} = \sqrt[4]{-16}$$

Step 2. −16 is negative and 4 is even, so $(-16)^{1/4}$ is not a real number.

$(-16)^{1/4} = \sqrt[4]{-16}$ is not defined for real numbers.

$(-16)^{1/4} \neq -2.$
$-2 \cdot -2 \cdot -2 \cdot -2 = 16$, not −16.

d. $\left(-\frac{32}{243}\right)^{1/5}$

Step 1. Apply $x^{1/n} = \sqrt[n]{x}$.

$$\left(-\frac{32}{243}\right)^{1/5} = \sqrt[5]{-\frac{32}{243}}$$

Step 2. Find the principal fifth root of $-\frac{32}{243}$.

$$\left(-\frac{32}{243}\right)^{1/5} = \sqrt[5]{-\frac{32}{243}} = -\frac{2}{3}$$

When you evaluate exponential expressions that have unit fraction exponents, you should practice doing Step 1 mentally. For instance, $49^{1/2} = 7$, $(-8)^{1/3} = -2$, $\left(\frac{1}{32}\right)^{1/5} = \frac{1}{2}$, and so forth.

Rational Exponents

If x is a real number and m and n are natural numbers, then (a) $x^{m/n} = (x^{1/n})^m$ or (b) $x^{m/n} = (x^m)^{1/n}$, provided that in all cases even roots of negative numbers do not occur.

When you evaluate the exponential expression $x^{m/n}$, you can find the nth root of x first and then raise the result to the mth power, or you can raise x to the mth power first and then find the nth root of the result. For most numerical situations, you usually will find it easier to find the root first and then raise to the power (as you will observe from the sample problems shown here).

Problem Evaluate using $x^{m/n} = (x^{1/n})^m$.

a. $\left(-\frac{1}{8}\right)^{2/3}$

b. $(36)^{3/2}$

Solution

a. $\left(-\frac{1}{8}\right)^{2/3}$

Step 1. Rewrite $\left(-\frac{1}{8}\right)^{2/3}$ using $x^{m/n}=(x^{1/n})^m$.

$$\left(-\frac{1}{8}\right)^{2/3}=\left[\left(-\frac{1}{8}\right)^{1/3}\right]^2$$

Step 2. Find $\left(-\frac{1}{8}\right)^{1/3}$.

$$\left(-\frac{1}{8}\right)^{1/3}=-\frac{1}{2}$$

Step 3. Raise $-\frac{1}{2}$ to the second power.

$$\left[-\frac{1}{2}\right]^2=\frac{1}{4}$$

Step 4. Review the main results.

$$\left(-\frac{1}{8}\right)^{2/3}=\left[\left(-\frac{1}{8}\right)^{1/3}\right]^2=\left[-\frac{1}{2}\right]^2=\frac{1}{4}$$

b. $(36)^{3/2}$

Step 1. Rewrite $(36)^{3/2}$ using $x^{m/n}=(x^{1/n})^m$.

$$(36)^{3/2}=(36^{1/2})^3$$

Step 2. Find $(36)^{1/2}$.

$$(36)^{1/2}=6$$

Step 3. Raise 6 to the third power.

$$6^3=216$$

Step 4. Review the main results.

$$(36)^{3/2}=(36^{1/2})^3=6^3=216$$

> $(36)^{3/2}\neq 36\cdot\frac{3}{2}$. $(36)^{3/2}=216$, but $36\cdot\frac{3}{2}=54$. Don't multiply the base by the exponent!

Problem Evaluate using $x^{m/n} = (x^m)^{1/n}$.

a. $\left(-\frac{1}{8}\right)^{2/3}$

b. $(36)^{3/2}$

Solution

a. $\left(-\frac{1}{8}\right)^{2/3}$

Step 1. Rewrite $\left(-\frac{1}{8}\right)^{2/3}$ using $x^{m/n} = (x^m)^{1/n}$.

$$\left(-\frac{1}{8}\right)^{2/3} = \left[\left(-\frac{1}{8}\right)^2\right]^{1/3}$$

Step 2. Find $\left(-\frac{1}{8}\right)^2$.

$$\left(-\frac{1}{8}\right)^2 = \frac{1}{64}$$

Step 3. Find $\left[\frac{1}{64}\right]^{1/3}$.

$$\left[\frac{1}{64}\right]^{1/3} = \frac{1}{4}$$

Step 4. Review the main results.

$$\left(-\frac{1}{8}\right)^{2/3} = \left[\left(-\frac{1}{8}\right)^2\right]^{1/3} = \left[\frac{1}{64}\right]^{1/3} = \frac{1}{4}$$

b. $(36)^{3/2}$

Step 1. Rewrite $(36)^{3/2}$ using $x^{m/n} = (x^m)^{1/n}$.

$$(36)^{3/2} = (36^3)^{1/2}$$

Step 2. Find $(36)^3$.

$$(36)^3 = 46{,}656$$

Step 3. Find $(46{,}656)^{1/2}$.

$$(46{,}656)^{1/2} = 216$$

Step 4. Review the main results.

$$(36)^{3/2} = \left(36^3\right)^{1/2} = (46{,}656)^{1/2} = 216$$

Exercise 4

For 1 and 2, write the indicated product as an exponential expression.

1. $-4 \cdot -4 \cdot -4 \cdot -4 \cdot -4$
2. $8 \cdot 8 \cdot 8 \cdot 8 \cdot 8 \cdot 8 \cdot 8$

For 3–18, evaluate, if possible.

3. $(-2)^7$
4. $(0.3)^4$
5. $\left(-\frac{3}{4}\right)^2$
6. 0^9
7. $(1+1)^5$
8. $(-2)^0$
9. 3^{-4}
10. $(-4)^{-2}$
11. $(0.3)^{-2}$
12. $\left(\frac{3}{4}\right)^{-1}$
13. $(-125)^{1/3}$
14. $(0.16)^{1/2}$
15. $(-121)^{1/4}$
16. $\left(\frac{16}{625}\right)^{1/4}$
17. $(-27)^{2/3}$
18. $\left(\frac{16}{625}\right)^{3/4}$

For 19 and 20, simplify.

19. $\frac{1}{5^{-3}}$
20. $\frac{1}{(-2)^{-4}}$

5

Order of Operations

In this chapter, you apply your skills in computation to perform a series of indicated numerical operations. This chapter lays the foundation for numerical calculations by introducing you to the order of operations.

Grouping Symbols

Grouping symbols such as parentheses ( ), brackets [ ], and braces { } are used to keep things together that belong together.

Fraction bars, absolute value bars | |, and square root symbols $\sqrt{\ }$ are also grouping symbols. When you are performing computations, perform operations in grouping symbols first.

Do keep in mind that parentheses are also used to indicate multiplication, as in (–5)(–8) or for clarity, as in –(–35).

Grouping symbols say "Do me first!"

It is *very important* that you do so when you have addition or subtraction inside the grouping symbol.

Problem Simplify.

a. $(1+1)^4$

b. $\dfrac{4+10}{4}$

c. $\dfrac{-7+25}{3-5}$

d. $|8+-5|$

e. $\sqrt{36+64}$

Solution

a. $(1+1)^4$

Step 1. Parentheses are a grouping symbol, so do $1+1$ *first*.

$$(1+1)^4 = 2^4$$

When you no longer need the grouping symbol, omit it.

Step 2. Evaluate 2^4.

$$=16$$

$(1+1)^4 \neq 1^4+1^4$. $(1+1)^4 = 16$, but $1^4+1^4 = 1+1 = 2$. Not performing the addition, $1+1$, inside the parentheses *first* can lead to an incorrect result.

b. $\frac{4+10}{4}$

Step 1. The fraction bar is a grouping symbol, so do the addition, $4+10$, over the fraction bar *first*.

$$\frac{4+10}{4} = \frac{14}{4}$$

$\frac{4+10}{4} \neq \frac{\cancel{4}+10}{\cancel{4}} \neq \frac{10}{1}$. Not performing the addition, $4+10$, *first* can lead to an incorrect result.

Step 2. Simplify $\frac{14}{4}$.

$$=\frac{7}{2}$$

c. $\frac{-7+25}{3-5}$

Step 1. The fraction bar is a grouping symbol, so do the addition, $-7+25$, over the fraction bar and the subtraction, $3-5$, under the fraction bar *first*.

$$\frac{-7+25}{3-5} = \frac{18}{-2}$$

$\frac{-7+25}{3-5} \neq \frac{-7+\cancel{25}^{5}}{3-\cancel{5}_{1}} \neq \frac{-7+5}{3-1}$. $\frac{-7+25}{3-5} = -9$ but $\frac{-7+5}{3-1} = \frac{-2}{2} = -1$. Not performing the addition, $-7+25$, and the subtraction, $3-5$, *first* can lead to an incorrect result.

Step 2. Compute $\frac{18}{-2}$.

$$=-9$$

d. $|8+-15|$

Step 1. Absolute value bars are a grouping symbol, so do $8+-15$ *first.*

$|8+-15| = |-7|$

Step 2. Evaluate $|-7|$.

$= 7$

> $|8+-15| \neq |8|+|-15|$. $|8+-15| = 7$, but $|8|+|-15| = 8+15 = 23$. Not performing the addition, $8+-15$, *first* can lead to an incorrect result.

e. $\sqrt{36+64}$

Step 1. The square root symbol is a grouping symbol, so do $36+64$ *first.*

$\sqrt{36+64} = \sqrt{100}$

Step 2. Evaluate $\sqrt{100}$.

$= 10$

> $\sqrt{36+64} \neq \sqrt{36}+\sqrt{64}$. $\sqrt{36+64} = 10$, $\sqrt{36}+\sqrt{64} = 6+8 = 14$. Not performing the addition, $36+64$, *first* can lead to an incorrect result.

PEMDAS

You must follow the order of operations to simplify mathematical expressions. Use the mnemonic "**P**lease **E**xcuse **M**y **D**ear **A**unt **S**ally"—abbreviated as PE(MD)(AS) to help you remember the following order.

Order of Operations

1. Do computations inside **P**arentheses (or other grouping symbols).
2. Evaluate **E**xponential expressions (also, evaluate absolute value, square root, and other root expressions).
3. Perform **M**ultiplication and **D**ivision, in the order in which these operations occur from left to right.
4. Perform **A**ddition and **S**ubtraction, in the order in which these operations occur from left to right.

> In the order of operations, multiplication does not always have to be done before division, or addition before subtraction. You multiply and divide in the order they occur in the problem. Similarly, you add and subtract in the order they occur in the problem.

Problem Simplify.

a. $\frac{60}{12} - 3 \cdot 4 + (1+1)^3$

b. $100 + 8 \cdot 3^2 - 63 \div (2+5)$

c. $\frac{-7+25}{3-5} + |8 + -15| - (5-3)^3$

Solution

a. $\frac{60}{12} - 3 \cdot 4 + (1+1)^3$

Step 1. Compute $1+1$ inside the parentheses.

$$\frac{60}{12} - 3 \cdot 4 + (1+1)^3$$

$$= \frac{60}{12} - 3 \cdot 4 + \mathbf{2}^3$$

Step 2. Evaluate 2^3.

$$= \frac{60}{12} - 3 \cdot 4 + \mathbf{8}$$

Step 3. Compute $\frac{60}{12}$.

$$= \mathbf{5} - 3 \cdot 4 + 8$$

$5 - 3 \cdot 4 + 8 \neq 2 \cdot 12$. Multiply *before* adding or subtracting—when no grouping symbols are present.

Step 4. Compute $3 \cdot 4$.

$$= 5 - \mathbf{12} + 8$$

Step 5. Compute $5 - 12$.

$$= \mathbf{-7} + 8$$

Step 6. Compute $-7 + 8$.

$$= \mathbf{1}$$

Step 7. Review the main steps.

$$\frac{60}{12} - 3 \cdot 4 + (1+1)^3 = \frac{60}{12} - 3 \cdot 4 + 2^3 = \frac{60}{12} - 3 \cdot 4 + 8 = 5 - 12 + 8 = 1$$

b. $100+8\cdot3^2-63\div(2+5)$

Step 1. Compute $2+5$ inside the parentheses.

$$100+8\cdot3^2-63\div(2+5)$$

$$=100+8\cdot3^2-63\div\mathbf{7}$$

> $8\cdot3^2\neq24^2$. $8\cdot3^2=8\cdot9=72$, but $24^2=576$. Do exponentiation *before* multiplication.

Step 2. Evaluate 3^2.

$$=100+8\cdot\mathbf{9}-63\div7$$

> $100+8\cdot9\neq108\cdot9$. Do multiplication *before* addition (except when a grouping symbol indicates otherwise).

Step 3. Compute $8\cdot9$.

$$=100+\mathbf{72}-63\div7$$

> $72-63\div7\neq9\div7$. Do division *before* subtraction (except when a grouping symbol indicates otherwise).

Step 4. Compute $63\div7$.

$$=100+72-\mathbf{9}$$

Step 5. Compute $100+72$.

$$=\mathbf{172}-9$$

Step 6. Compute $172-9$.

$$=\mathbf{163}$$

Step 7. Review the main steps.

$$100+8\cdot3^2-63\div(2+5)=100+8\cdot3^2-63\div7=100+8\cdot9-63\div7$$
$$=100+72-9=163$$

c. $\dfrac{-7+25}{3-5}+|8+-15|-(5-3)^3$

Step 1. Compute quantities in grouping symbols.

$$\frac{-7+25}{3-5}+|8+-15|-(5-3)^3$$

$$=\frac{\mathbf{18}}{\mathbf{-2}}+|\mathbf{-7}|-\mathbf{2}^3$$

Step 2. Evaluate $|-7|$ and 2^3.

$$=\frac{18}{-2}+\mathbf{7}-\mathbf{8}$$

> Evaluate absolute value expressions *before* multiplication or division.

Step 3. Compute $\frac{18}{-2}$.

$= \mathbf{-9} + 7 - 8$

Step 4. Compute $-9+7$.

$= \mathbf{-2} - 8$

Step 5. Compute $-2-8$.

$= \mathbf{-10}$

Step 6. Review the main steps.

$$\frac{-7+25}{3-5} + |8+-15| - (5-3)^3 = \frac{18}{-2} + |-7| - 2^3 = \frac{18}{-2} + 7 - 8 =$$
$$-9+7-8 = -10$$

Exercise 5

Simplify.

1. $(5+7)6-10$
2. $(-7^2)(6-8)$
3. $(2-3)(-20)$
4. $3(-2) - \frac{10}{-5}$
5. $9 - \frac{20+22}{6} - 2^3$
6. $-2^2 \cdot -3 - (15-4)^2$
7. $5(11-3-6\cdot 2)^2$
8. $-10 - \frac{-8-(3\cdot -3+15)}{2}$
9. $\frac{7^2 - 8\cdot 5 + 3^4}{3\cdot 2 - 36 \div 12}$
10. $(-6)\left(\frac{\sqrt{625-576}}{14}\right) + \frac{6}{-3}$
11. $\frac{5-|-5|}{20^2}$
12. $(12-5)-(5-12)$
13. $\frac{9+\sqrt{100-64}}{-|-15|}$
14. $-8+2(-1)^2+6$
15. $\frac{3}{2}\left(-\frac{2}{3}\right) - \frac{1}{4}(-5) + \frac{15}{7}\left(-\frac{7}{3}\right)$

6

Algebraic Expressions

This chapter presents a discussion of algebraic expressions. It begins with the basic terminology that is critical to your understanding of the concept of an algebraic expression.

Algebraic Terminology

A *variable* holds a place open for a number (or numbers, in some cases) whose value may vary. You usually express a variable as an upper or lowercase letter (e.g., x, y, z, A, B, or C); for simplicity, the letter is the "name" of the variable. In problem situations, you use variables to represent unknown quantities. Although a variable may represent any number, in many problems the variables represent specific numbers, but the values are unknown.

> You can think of variables as numbers in disguise. Not recognizing that variables represent numbers is a common mistake for beginning students of algebra.

A *constant* is a quantity that has a fixed, definite value that does not change in a problem situation. For example, all the real numbers are constants, including πe real numbers whose units are units of measure such as 5 feet, 60 degrees, 100 pounds, and so forth.

Problem Name the variable(s) and constant(s) in the given expression.

a. $\frac{5}{9}(F-32)$, where F is the number of degrees Fahrenheit

b. πd, where d is the measure of the diameter of a circle

Solution

a. $\frac{5}{9}(F - 32)$, where F is the number of degrees Fahrenheit

Step 1. Recall that a letter names a variable whose value may vary.

Step 2. Name the variable(s).

The letter F stands for the number of degrees Fahrenheit and can be any real number, and so it is a variable.

Step 3. Recall that a constant has a fixed, definite value.

Step 4. Name the constant(s).

The numbers $\frac{5}{9}$ and 32 have fixed, definite values that do not change, and so they are constants.

b. πd, where d is the measure of the diameter of a circle

Step 1. Recall that a letter names a variable whose value may vary.

Step 2. Name the variable(s).

The letter d stands for the measure of the diameter of a circle and can be any nonnegative number, and so it is a variable.

Step 3. Recall that a constant has a fixed, definite value.

Step 4. Name the constant(s).

The number π has a fixed, definite value that does not change, and so it is a constant.

> Even though the number pi is represented by a Greek letter, π is not a variable. The number π is an irrational number whose approximate value to two decimal places is 3.14.

If there is a number immediately next to a variable (normally preceding it), that number is the *numerical coefficient* of the variable. If there is no number written immediately next to a variable, it is understood that the numerical coefficient is 1.

Problem What is the numerical coefficient of the variable?

a. $-5x$

b. y

c. πd

Solution

a. $-5x$

Step 1. Identify the numerical coefficient by observing that the number -5 immediately precedes the variable x.

-5 is the numerical coefficient of x.

b. y

Step 1. Identify the numerical coefficient by observing that no number is written immediately next to the variable y.

1 is the numerical coefficient of y.

c. πd

Step 1. Identify the numerical coefficient by observing that the number π immediately precedes the variable d.

π is the numerical coefficient of d.

Evaluating Algebraic Expressions

Writing variables and coefficients or two or more variables (with or without constants) side by side with no multiplication symbol in between is a way to show multiplication. Thus, $-5x$ means -5 times x, and $2xyz$ means 2 times x times y times z. Also, a number or variable written immediately next to a grouping symbol indicates multiplication. For instance, $6(x+1)$ means 6 times the quantity $(x+1)$, $7\sqrt{x}$ means 7 times $\sqrt{x}$, and $-1|-8|$ means -1 times $|-8|$.

An *algebraic expression* is a symbolic representation of a number. It can contain constants, variables, and computation symbols. Here are examples of algebraic expressions.

$$-5x,\ 2xyz,\ \frac{6(x+1)}{7\sqrt{x}+1},\ \frac{-1|y|+5(x-y)}{z+1},\ -8xy^3+\frac{5}{2x^2}-27,\ 8a^3+64b^3,$$

$$x^2-x-12,\ \frac{1}{3}x^2z^3,\ \text{and } -2x^5+5x^4-3x^3-7x^2+x+4$$

Ordinarily, you don't know what number an algebraic expression represents because algebraic expressions always contain variables. However, if you are given numerical values for the variables, you can evaluate the algebraic expression by substituting the given numerical value for each variable and then simplifying by performing the indicated operations, being sure to *follow the order of operations* as you proceed.

Problem Find the value of the algebraic expression when $x = 4$, $y = -8$, and $z = -5$.

a. $-5x$

b. $2xyz$

c. $\dfrac{6(x+1)}{7\sqrt{x}+1}$

d. $\dfrac{-1|y|+5(x-y)}{z+1}$

e. $x^2 - x - 12$

Solution

a. $-5x$

Step 1. Substitute 4 for x in the expression $-5x$.

$$-5x = -5(4)$$

Step 2. Perform the indicated multiplication.

$$= -20$$

Step 3. State the main result.

$-5x = -20$ when $x = 4$.

b. $2xyz$

When you substitute negative values into an algebraic expression, enclose them in parentheses to avoid careless errors.

Step 1. Substitute 4 for x, -8 for y, and -5 for z in the expression $2xyz$.

$$2xyz = 2(4)(-8)(-5)$$

Step 2. Perform the indicated multiplication.

$$= 320$$

Step 3. State the main result.

$2xyz = 320$ when $x = 4$, $y = -8$, and $z = -5$.

c. $\dfrac{6(x+1)}{7\sqrt{x}+1}$

Step 1. Substitute 4 for x in the expression $\dfrac{6(x+1)}{7\sqrt{x}+1}$.

$$\frac{6(x+1)}{7\sqrt{x}+1} = \frac{6(4+1)}{7\sqrt{4}+1}$$

$7\sqrt{4}+1 \neq 7\sqrt{5}$. The square root applies only to the 4.

Step 2. Simplify the resulting expression.

$$= \frac{6(5)}{7 \cdot 2+1}$$

$$= \frac{30}{14+1}$$

$$= \frac{30}{15}$$

$$= 2$$

Step 3. State the main result.

$\dfrac{6(x+1)}{7\sqrt{x}+1} = 2$ when $x = 4$.

d. $\dfrac{-1|y|+5(x-y)}{z+1}$

Step 1. Substitute 4 for x, –8 for y, and –5 for z in the expression $\dfrac{-1|y|+5(x-y)}{z+1}$.

$$\frac{-1|y|+5(x-y)}{z+1} = \frac{-1|-8|+5(4-(-8))}{(-5)+1}$$

Step 2. Simplify the resulting expression.

$$= \frac{-1(8)+5(4+8)}{-5+1}$$

$$= \frac{-8+5(12)}{-4}$$

$$= \frac{-8+60}{-4}$$

$$= \frac{52}{-4}$$

$$= -13$$

Step 3. State the main result.

$$\frac{-1|y| + 5(x - y)}{z + 1} = -13 \text{ when } x = 4,\ y = -8, \text{ and } z = -5.$$

e. $x^2 - x - 12$

Step 1. Substitute 4 for x in the expression $x^2 - x - 12$.

$$x^2 - x - 12 = (4)^2 - (4) - 12$$

Step 2. Simplify the resulting expression.

$$= (4)^2 - (4) - 12$$

$$= 16 - 4 - 12$$

$$= 0$$

Step 3. State the main result.

$$x^2 - x - 12 = 0 \text{ when } x = 4.$$

Problem Evaluate $-2x^5 + 5x^4 - 3x^3 - 7x^2 + x + 4$ when $x = -1$.

Solution

Step 1. Substitute $x = -1$ for x in the expression $-2x^5 + 5x^4 - 3x^3 - 7x^2 + x + 4$.

$$-2x^5 + 5x^4 - 3x^3 - 7x^2 + x + 4$$

$$= -2(-1)^5 + 5(-1)^4 - 3(-1)^3 - 7(-1)^2 + (-1) + 4$$

Step 2. Simplify the resulting expression.

$$= 2 + 5 + 3 - 7 - 1 + 4$$

$$= 6$$

> Watch your signs! It's easy to make careless errors when you are evaluating negative numbers raised to powers.

Step 3. State the main result.

$$-2x^5 + 5x^4 - 3x^3 - 7x^2 + x + 4 = 6 \text{ when } x = -1.$$

You can use your skills in evaluating algebraic expressions to evaluate formulas for given numerical values.

Problem Find C when $F = 212$ using the formula $C = \frac{5}{9}(F - 32)$.

Solution

Step 1. Substitute 212 for F in the formula $C = \frac{5}{9}(F - 32)$.

$$C = \frac{5}{9}(F - 32)$$

$$C = \frac{5}{9}(212 - 32)$$

Step 2. Simplify.

$$C = \frac{5}{9}(212 - 32)$$

$$C = \frac{5}{9}(180)$$

$$C = 100$$

Step 3. State the main result.

$C = 100$ when $F = 212$.

Dealing with Parentheses

Frequently, algebraic expressions are enclosed in parentheses. It is important that you deal with parentheses correctly.

If no symbol or if a (+) symbol immediately precedes parentheses that enclose an algebraic expression, remove the parentheses and rewrite the algebraic expression without changing any signs.

Problem Simplify $\left(-8xy^3 + \frac{5}{2x^2} - 27\right)$.

Solution

Step 1. Remove the parentheses without changing any signs.

$$\left(-8xy^3 + \frac{5}{2x^2} - 27\right) = -8xy^3 + \frac{5}{2x^2} - 27$$

If an opposite (–) symbol immediately precedes parentheses that enclose an algebraic expression, remove the parentheses and the opposite symbol and rewrite the algebraic expression, but with all the signs changed.

Problem Simplify $-\left(-8xy^3 + \dfrac{5}{2x^2} - 27\right)$.

Solution

Step 1. Remove the parentheses and the opposite symbol and rewrite the expression, but change all the signs.

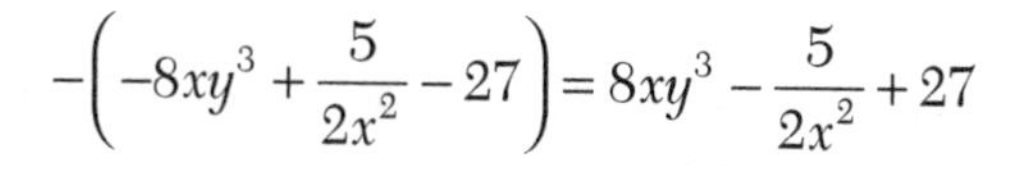

$$-\left(-8xy^3 + \frac{5}{2x^2} - 27\right) = 8xy^3 - \frac{5}{2x^2} + 27$$

$$-\left(-8xy^3 + \frac{5}{2x^2} - 27\right) \neq 8xy^3 + \frac{5}{2x^2} - 27.$$

Change *all* the signs, not just the first one. This mistake is very common.

If a minus (–) symbol immediately precedes parentheses that enclose an algebraic expression, mentally think of the minus symbol as "+–," meaning "add the opposite." Then remove the parentheses and rewrite the algebraic expression, but change all the signs.

Problem Simplify $10 - \left(3x^3 - 7x^2 + 2x\right)$.

Solution

Step 1. Mentally think of the minus symbol as "+–."

$$\underbrace{10 + -\left(3x^3 - 7x^2 + 2x\right)}_{\text{Do this mentally.}}$$

Step 2. Remove the parentheses and rewrite the algebraic expression, but with all the signs changed.

$$10 - \left(3x^3 - 7x^2 + 2x\right) = 10 - 3x^3 + 7x^2 - 2x$$

If a number immediately precedes parentheses that enclose an algebraic expression, apply the distributive property to remove the parentheses.

Problem Simplify $2(x+5)$.

Solution

Step 1. Apply the distributive property.

$$2(x+5)$$
$$=2\cdot x+2\cdot 5$$
$$=2x+10$$

> $2(x+5)\neq 2x+5$. You must multiply the 5 by 2 as well.

Exercise 6

1. Name the variable(s) and constant(s) in the expression $2\pi r$, where r is the measure of the radius of a circle.

For 2–4, state the numerical coefficient of the variable.

2. $-12y$
3. z
4. $\frac{2}{3}x$

For 5–12, find the value of the algebraic expression when $x = 9$, $y = -2$, and $z = -3$.

5. $-5x$
6. $2xyz$
7. $\dfrac{6(x+1)}{5\sqrt{x}-10}$
8. $\dfrac{-2|y|+5(2x-y)}{-6z+y^3}$
9. x^2-8x-9
10. $2y+x(y-z)$
11. $\dfrac{(x+y)^2}{x^2-y^2}$
12. $(y+z)^{-3}$

For 13–15, find the variable using the formula given.

13. Find A when $b = 12$ and $h = 8$ using the formula $A=\frac{1}{2}bh$.
14. Find V when $r = 5$ and $h = 18$ using the formula $V=\frac{1}{3}\pi r^2h$. Use $\pi = 3.14$.
15. Find c when $a = 8$ and $b = 15$ using the formula $c^2=a^2+b^2$.

For 16–20, simplify by removing parentheses.

16. $-\left(-\frac{1}{2}x^3y^2 + 7xy^3 - 30\right)$

17. $\left(8a^3 + 64b^3\right)$

18. $-4 - \left(-2y^3\right)$

19. $-3(x + 4)$

20. $12 + \left(x^2 + y\right)$

7

Rules for Exponents

In Chapter 4, you learned about the various types of exponents that you might encounter in algebra. In this chapter, you learn about the rules for exponents—which you will find useful when you simplify algebraic expressions. The following rules hold for all real numbers x and y and all rational numbers m, n, and p, provided that all indicated powers are real and no denominator is 0.

Product Rule

Product Rule for Exponential Expressions with the Same Base

$$x^m x^n = x^{m+n}$$

This rule tells you that when you multiply exponential expressions that have the *same* base, you *add* the exponents and keep the same base.

> If the bases are not the same, don't use the product rule for exponential expressions with the same base.

Problem Simplify.

a. x^2x^3

b. x^2y^5

c. $x^2x^7y^3y^5$

Solution

a. x^2x^3

Step 1. Check for exponential expressions that have the same base.

x^2x^3

x^2 and x^3 have the same base, namely, x.

Step 2. Simplify x^2x^3. Keep the base x and add the exponents 2 and 3.

$x^2x^3 = x^{2+3} = x^5$

$x^2x^3 \neq x^{2\cdot3} = x^6$. When multiplying, *add* the exponents of the same base, don't multiply them.

b. x^2y^5

Step 1. Check for exponential expressions that have the same base.

x^2y^5

x^2 and y^5 do not have the same base, so the product cannot be simplified.

$x^2y^5 \neq (xy)^7$. This is a common error that you should avoid.

c. $x^2x^7y^3y^5$

Step 1. Check for exponential expressions that have the same base.

$x^2x^7y^3y^5$

x^2 and x^7 have the same base, namely, x, and y^3 and y^5 have the same base, namely, y.

Step 2. Simplify x^2x^7 and y^3y^5. For each, keep the base and add the exponents.

$x^2x^7y^3y^5 = x^{2+7}y^{3+5} = x^9y^8$

Quotient Rule

Quotient Rule for Exponential Expressions with the Same Base

$$\frac{x^m}{x^n} = x^{m-n},\ x \neq 0$$

This rule tells you that when you divide exponential expressions that have the *same* base, you *subtract* the denominator exponent from the numerator exponent and keep the same base.

If the bases are not the same, don't use the quotient rule for exponential expressions with the same base.

Problem Simplify.

a. $\frac{x^5}{x^3}$

b. $\frac{y^5}{x^2}$

c. $\frac{x^7y^5}{x^2y^3}$

d. $\frac{x^3}{x^{10}}$

Solution

a. $\frac{x^5}{x^3}$

Step 1. Check for exponential expressions that have the same base.

$$\frac{x^5}{x^3}$$

x^5 and x^3 have the same base, namely, x.

Step 2. Simplify $\frac{x^5}{x^3}$. Keep the base x and subtract the exponents 5 and 3.

$$\frac{x^5}{x^3} = x^{5-3} = x^2$$

$\frac{x^5}{x^3} \neq x^{5/3}$. When dividing, *subtract* the exponents of the same base, don't divide them.

b. $\frac{y^5}{x^2}$

Step 1. Check for exponential expressions that have the same base.

$$\frac{y^5}{x^2}$$

$\frac{y^5}{x^2} \neq \left(\frac{y}{x}\right)^3$. This is a common error that you should avoid.

y^5 and x^2 do not have the same base, so the quotient cannot be simplified.

c. $\dfrac{x^7y^5}{x^2y^3}$

Step 1. Check for exponential expressions that have the same base.

$$\frac{x^7y^5}{x^2y^3}$$

x^7 and x^2 have the same base, namely, x, and y^5 and y^3 have the same base, namely, y.

Step 2. Simplify $\dfrac{x^7}{x^2}$ and $\dfrac{y^5}{y^3}$. For each, keep the base and subtract the exponents.

$$\frac{x^7y^5}{x^2y^3} = x^{7-2}y^{5-3} = x^5y^2$$

d. $\dfrac{x^3}{x^{10}}$

Step 1. Check for exponential expressions that have the same base.

$$\frac{x^3}{x^{10}}$$

x^3 and x^{10} have the same base, namely, x.

Step 2. Simplify $\dfrac{x^3}{x^{10}}$. Keep the base x and subtract the exponents 3 and 10.

$$\frac{x^3}{x^{10}} = x^{3-10} = x^{-7}$$

Step 3. Express x^{-7} as an equivalent exponential expression with a positive exponent.

$$x^{-7} = \frac{1}{x^7}$$

When you simplify expressions, make sure your final answer does not contain negative exponents.

Rules for Powers

Rule for a Power to a Power

$$(x^m)^p = x^{mp}$$

This rule tells you that when you raise an exponential expression to a power, keep the base and *multiply* exponents.

Problem Simplify.

a. $(x^2)^3$

b. $(y^3)^5$

Solution

a. $(x^2)^3$

Step 1. Keep the same base x and multiply the exponents 2 and 3.

$$(x^2)^3 = x^{2\cdot3} = x^6$$

$(x^2)^3 \neq x^5$. For a power to a power, *multiply* exponents, don't add.

b. $(y^3)^5$

Step 1. Keep the same base y and multiply the exponents 3 and 5.

$$(y^3)^5 = y^{3\cdot5} = y^{15}$$

Rule for the Power of a Product

$$(xy)^p = x^p y^p$$

This rule tells you that a product raised to a power is the product of each factor raised to the power.

Confusing the rule $(xy)^p = x^p y^p$ with the rule $x^m x^n = x^{m+n}$ is a common error. Notice that the rule $(xy)^p = x^p y^p$ has the same exponent and different bases, while the rule $x^m x^n = x^{m+n}$ has the same base and different exponents.

Problem Simplify.

a. $(xy)^6$

b. $(4x)^3$

c. $(x^3y^2z)^4$

Solution

a. $(xy)^6$

Step 1. Raise each factor to the power of 6.

$$(xy)^6 = x^6y^6$$

b. $(4x)^3$

Step 1. Raise each factor to the power of 3.

$$(4x)^3 = 4^3x^3 = 64x^3$$

c. $(x^3y^2z)^4$

Step 1. Raise each factor to the power of 4.

$$(x^3y^2z)^4 = (x^3)^4(y^2)^4(z)^4 = x^{12}y^8z^4$$

Rule for the Power of a Quotient

$$\left(\frac{x}{y}\right)^p = \frac{x^p}{y^p},\ y \neq 0$$

This rule tells you that a quotient raised to a power is the quotient of each factor raised to the power.

Problem Simplify.

a. $\left(\frac{3}{x}\right)^4$

b. $\left(\frac{-4x}{5y}\right)^3$

Solution

a. $\left(\frac{3}{x}\right)^4$

Step 1. Raise each factor to the power of 4.

$$\left(\frac{3}{x}\right)^4 = \frac{(3)^4}{(x)^4} = \frac{81}{x^4}$$

b. $\left(\frac{-4x}{5y}\right)^3$

Step 1. Raise each factor to the power of 3.

$$\left(\frac{-4x}{5y}\right)^3 = \frac{(-4x)^3}{(5y)^3} = \frac{(-4)^3(x)^3}{(5)^3(y)^3} = \frac{-64x^3}{125y^3} = -\frac{64x^3}{125y^3}$$

Rules for Exponents Summary

You must be very careful when simplifying using rules for exponents. For your convenience, here is a summary of the rules.

Rules for Exponents

1. **Product Rule for Exponential Expressions with the Same Base**

   $$x^m x^n = x^{m+n}$$

2. **Quotient Rule for Exponential Expressions with the Same Base**

   $$\frac{x^m}{x^n} = x^{m-n},\ x \neq 0$$

3. **Rule for a Power to a Power**

   $$(x^m)^p = x^{mp}$$

4. **Rule for the Power of a Product**

   $$(xy)^p = x^p y^p$$

5. **Rule for the Power of a Quotient**

   $$\left(\frac{x}{y}\right)^p = \frac{x^p}{y^p},\ y \neq 0$$

Notice there is no rule for the power of a sum [e.g., $(x+y)^2$] or for the power of a difference [e.g., $(x-y)^2$]. Therefore, an algebraic sum or difference raised to a power cannot be simplified using only rules for exponents.

Problem Simplify using only rules for exponents.

a. $(xy)^2$

b. $(x+y)^2$

c. $(x-y)^2$

d. $(x+y)^2(x+y)^3$

e. $\dfrac{(x-y)^5}{(x-y)^2}$

Solution

a. $(xy)^2$

Step 1. This is a power of a product, so square each factor.

$$(xy)^2 = x^2y^2$$

b. $(x+y)^2$

Step 1. This is a power of a sum. It cannot be simplified using only rules for exponents.

$(x+y)^2$ is the answer.

> $(x+y)^2 \neq x^2+y^2$! $(x+y)^2=(x+y)(x+y)$ $=x^2+2xy+y^2$ (which you will learn in Chapter 9). This is the most common error that beginning algebra students make.

c. $(x-y)^2$

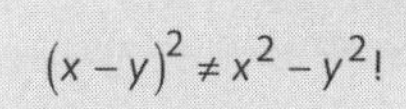

Step 1. This is a power of a difference. It cannot be simplified using only rules for exponents.

$(x-y)^2$ is the answer.

d. $(x+y)^2(x+y)^3$

Step 1. This is a product of expressions with the same base, namely, $(x+y)$. Keep the base and add the exponents.

$$(x+y)^2(x+y)^3=(x+y)^{2+3}=(x+y)^5$$

> When a quantity enclosed in a grouping symbol acts as a base, you can use the rules for exponents to simplify as long as you continue to treat the quantity as a base.

Step 2. $(x+y)^5$ is a power of a sum. It cannot be simplified using only rules for exponents.

$(x+y)^5$ is the answer.

e. $\dfrac{(x-y)^5}{(x-y)^2}$

Step 1. This is a quotient of expressions with the same base, namely, $(x-y)$. Keep the base and subtract the exponents.

$$\frac{(x-y)^5}{(x-y)^2}=(x-y)^{5-2}=(x-y)^3$$

Step 2. $(x-y)^3$ is a power of a difference. It cannot be simplified using only rules for exponents.

$(x-y)^3$ is the answer.

Exercise 7

Simplify using only rules for exponents.

1. x^4x^9
2. $x^3x^4y^6y^5$
3. $\dfrac{x^6}{x^3}$
4. $\dfrac{x^5y^5}{x^2y^4}$
5. $\dfrac{x^4}{x^6}$
6. $\left(x^2\right)^5$
7. $(xy)^5$
8. $(-5x)^3$
9. $\left(2x^5yz^3\right)^4$
10. $\left(\dfrac{5}{3x}\right)^4$
11. $\left(\dfrac{-3x}{5y}\right)^4$
12. $(2x+1)^2$
13. $(3x-5)^3$
14. $(x+3)(x+3)^2$
15. $\dfrac{(2x-y)^{15}}{(2x-y)^5}$

8

Adding and Subtracting Polynomials

In this chapter, you learn how to add and subtract polynomials. It begins with a discussion of the elementary concepts that you need to know to ensure your success when working with polynomials.

Terms and Monomials

In an algebraic expression, *terms* are the parts of the expression that are connected to the other parts by plus or minus symbols. If the algebraic expression has no plus or minus symbols, then the algebraic expression itself is a term.

Problem Identify the terms in the given expression.

a. $-8xy^3 + \frac{5}{2x^2} - 27$

b. $3x^5$

Solution

a. $-8xy^3 + \frac{5}{2x^2} - 27$

Step 1. The expression contains plus and minus symbols, so identify the quantities between the plus and minus symbols.

The terms are $-8xy^3$, $\frac{5}{2x^2}$, and 27.

b. $3x^5$

Step 1. There are no plus or minus symbols, so the expression is a term.

The term is $3x^5$.

A *monomial* is a special type of term that when simplified is a constant or a product of one or more variables raised to nonnegative integer powers, with or without an explicit coefficient.

In monomials, no variable divisors, negative exponents, or fractional exponents are allowed.

Problem Specify whether the term is a monomial. Explain your answer.

a. $-8xy^3$

b. $\frac{5}{2x^2}$

c. 0

d. $3x^5$

e. 27

f. $4x^{-3}y^2$

g. $6\sqrt{x}$

Solution

a. $-8xy^3$

Step 1. Check whether $-8xy^3$ meets the criteria for a monomial.

$-8xy^3$ is a term that is a product of variables raised to positive integer powers, with an explicit coefficient of -8, so it is a monomial.

b. $\frac{5}{2x^2}$

Step 1. Check whether $\frac{5}{2x^2}$ meets the criteria for a monomial.

$\frac{5}{2x^2}$ is a term, but it contains division by a variable, so it is not a monomial.

c. 0

Step 1. Check whether 0 meets the criteria for a monomial.

0 is a constant, so it is a monomial.

d. $3x^5$

Step 1. Check whether $3x^5$ meets the criteria for a monomial.

$3x^5$ is a term that is a product of one variable raised to a positive integer power, with an explicit coefficient of 3, so it is a monomial.

e. 27

Step 1. Check whether 27 meets the criteria for a monomial.

27 is a constant, so it is a monomial.

f. $4x^{-3}y^2$

Step 1. Check whether $4x^{-3}y^2$ meets the criteria for a monomial.

$4x^{-3}y^2$ contains a negative exponent, so it is not a monomial.

g. $6\sqrt{x}$

Step 1. Check whether $6\sqrt{x}$ meets the criteria for a monomial.

$6\sqrt{x} = 6x^{\frac{1}{2}}$ contains a fractional exponent, so it is not a monomial.

Polynomials

A *polynomial* is a single monomial or a sum of monomials. A polynomial that has exactly one term is a *monomial*. A polynomial that has exactly two terms is a *binomial*. A polynomial that has exactly three terms is a *trinomial*. A polynomial that has more than three terms is just a general polynomial.

Problem State the most specific name for the given polynomial.

a. $x^2 - 1$

b. $8a^3 + 64b^3$

c. $x^2 + 4x - 12$

d. $\frac{1}{3}x^2z^3$

e. $-2x^5 + 5x^4 - 3x^3 - 7x^2 + x + 4$

Solution

a. $x^2 - 1$

Step 1. Count the terms of the polynomial.

$x^2 - 1$ has exactly two terms.

Step 2. State the specific name.

$x^2 - 1$ is a binomial.

b. $8a^3 + 64b^3$

Step 1. Count the terms of the polynomial.

$8a^3 + 64b^3$ has exactly two terms.

Step 2. State the specific name.

$8a^3 + 64b^3$ is a binomial.

c. $x^2 + 4x - 12$

Step 1. Count the terms of the polynomial.

$x^2 + 4x - 12$ has exactly three terms.

Step 2. State the specific name.

$x^2 + 4x - 12$ is a trinomial.

d. $\frac{1}{3}x^2z^3$

Step 1. Count the terms of the polynomial.

$\frac{1}{3}x^2z^3$ has exactly one term.

Step 2. State the specific name.

$\frac{1}{3}x^2z^3$ is a monomial.

e. $-2x^5 + 5x^4 - 3x^3 - 7x^2 + x + 4$

Step 1. Count the terms of the polynomial.

$-2x^5 + 5x^4 - 3x^3 - 7x^2 + x + 4$ has exactly six terms.

Step 2. State the specific name.

$-2x^5 + 5x^4 - 3x^3 - 7x^2 + x + 4$ is a polynomial.

Like Terms

Monomials that are constants or that have exactly the same variable factors (i.e., the same letters with the same corresponding exponents) are *like terms*. Like terms are the same except, perhaps, for their coefficients.

Problem State whether the given monomials are like terms. Explain your answer.

a. $-10x$ and $25x$

b. $4x^2y^3$ and $-7x^3y^2$

c. 100 and 45

d. 25 and $25x$

Solution

a. $-10x$ and $25x$

Step 1. Check whether $-10x$ and $25x$ meet the criteria for like terms.

$-10x$ and $25x$ are like terms because they are exactly the same except for their numerical coefficients.

b. $4x^2y^3$ and $-7x^3y^2$

Step 1. Check whether $4x^2y^3$ and $-7x^3y^2$ meet the criteria for like terms.

$4x^2y^3$ and $-7x^3y^2$ are not like terms because the corresponding exponents on x and y are not the same.

c. 100 and 45

Step 1. Check whether 100 and 45 meet the criteria for like terms.

100 and 45 are like terms because they are both constants.

d. 25 and $25x$

Step 1. Check whether 25 and $25x$ meet the criteria for like terms.

25 and $25x$ are not like terms because they do not contain the same variable factors.

Finally, monomials that are not like terms are *unlike terms*.

Addition and Subtraction of Monomials

Because variables are standing in for real numbers, you can use the properties of real numbers to perform operations with polynomials.

Addition and Subtraction of Monomials

1. To add monomials that are like terms, add their numerical coefficients and use the sum as the coefficient of their common variable component.
2. To subtract monomials that are like terms, subtract their numerical coefficients and use the difference as the coefficient of their common variable component.
3. To add or subtract unlike terms, indicate the addition or subtraction.

Problem Simplify.

a. $-10x+25x$

b. $4x^2y^3-7x^3y^2$

c. $9x^2+3x^2-7x^2$

d. $25+25x$

e. $5x^2-7x^2$

Solution

a. $-10x+25x$

Step 1. Check for like terms.

$$-10x+25x$$

$-10x$ and $25x$ are like terms.

Step 2. Add the numerical coefficients.

$$-10+25=15$$

Step 3. Use the sum as the coefficient of x.

$$-10x+25x=15x$$

$-10x+25x \neq 15x^2$. In addition and subtraction, the exponents on the variables do not change.

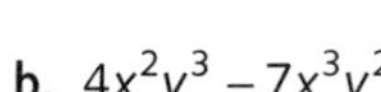
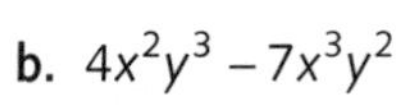

b. $4x^2y^3-7x^3y^2$

Step 1. Check for like terms.

$$4x^2y^3-7x^3y^2$$

$4x^2y^3$ and $7x^3y^2$ are not like terms, so leave the problem as indicated subtraction: $4x^2y^3 - 7x^3y^2$.

c. $9x^2 + 3x^2 - 7x^2$

Step 1. Check for like terms.

$9x^2 + 3x^2 - 7x^2$

$9x^2$, $3x^2$, and $7x^2$ are like terms.

Step 2. Combine the numerical coefficients.

$9 + 3 - 7 = 5$

Step 3. Use the result as the coefficient of x^2.

$9x^2 + 3x^2 - 7x^2 = 5x^2$

d. $25 + 25x$

Step 1. Check for like terms.

$25 + 25x$

25 and $25x$ are not like terms, so leave the problem as indicated addition: $25 + 25x$.

$25 + 25x \neq 50x$. These are not like terms, so you cannot combine them into one single term.

e. $5x^2 - 7x^2$

Step 1. Check for like terms.

$5x^2 - 7x^2$

$5x^2$ and $7x^2$ are like terms.

Step 2. Subtract the numerical coefficients.

$5 - 7 = -2$

Step 3. Use the result as the coefficient of x^2.

$5x^2 - 7x^2 = -2x^2$

Combining Like Terms

When you have an assortment of like terms in the same expression, systematically combine matching like terms in the expression. (For example, you might proceed from left to right.) To organize the process, use the properties of real numbers to rearrange the expression so that matching like terms are together (later, you might choose do this step mentally). If the expression includes unlike terms, just indicate the sums or differences of such terms. To avoid sign errors as you work, *keep a – symbol with the number that follows it.*

Problem Simplify $4x^3 + 5x^2 - 10x + 25 + 2x^3 - 7x^2 - 5$.

Solution

Step 1. Check for like terms.

$$4x^3 + 5x^2 - 10x + 25 + 2x^3 - 7x^2 - 5$$

The like terms are $4x^3$ and $2x^3$, $5x^2$ and $7x^2$, and 25 and 5.

Step 2. Rearrange the expression so that like terms are together.

$$4x^3 + 5x^2 - 10x + 25 + 2x^3 - 7x^2 - 5$$
$$= 4x^3 + 2x^3 + 5x^2 - 7x^2 - 10x + 25 - 5$$

When you are simplifying, rearrange so that like terms are together can be done mentally. However, writing out this step helps you avoid careless errors.

Step 3. Systematically combine matching like terms and indicate addition or subtraction of unlike terms.

$$= 6x^3 + -2x^2 - 10x + 20$$
$$= 6x^3 - 2x^2 - 10x + 20$$

Because + – is equivalent to –, it is customary to change + – to simply – when you are simplifying expressions.

Step 4. Review the main result.

$$4x^3 + 5x^2 - 10x + 25 + 2x^3 - 7x^2 - 5 = 6x^3 - 2x^2 - 10x + 20$$

Addition and Subtraction of Polynomials

Addition of Polynomials

To add two or more polynomials, add like monomial terms and simply indicate addition or subtraction of unlike terms.

Problem Perform the indicated addition.

a. $(9x^2 - 6x + 2) + (-7x^2 - 5x + 3)$

b. $(4x^3 + 3x^2 - x + 8) + (8x^3 + 2x - 10)$

Solution

a. $(9x^2 - 6x + 2) + (-7x^2 - 5x + 3)$

Step 1. Remove parentheses.

$$(9x^2 - 6x + 2) + (-7x^2 - 5x + 3)$$
$$= 9x^2 - 6x + 2 - 7x^2 - 5x + 3$$

Step 2. Rearrange the terms so that like terms are together. (You might do this step mentally.)

$$= 9x^2 - 7x^2 - 6x - 5x + 2 + 3$$

Step 3. Systematically combine matching like terms and indicate addition or subtraction of unlike terms.

$$= 2x^2 - 11x + 5$$

You should write polynomial answers in descending powers of a variable.

Step 4. Review the main results.

$$(9x^2 - 6x + 2) + (-7x^2 - 5x + 3)$$
$$= 9x^2 - 6x + 2 - 7x^2 - 5x + 3 = 2x^2 - 11x + 5$$

b. $(4x^3 + 3x^2 - x + 8) + (8x^3 + 2x - 10)$

Step 1. Remove parentheses.

$$(4x^3 + 3x^2 - x + 8) + (8x^3 + 2x - 10)$$
$$= 4x^3 + 3x^2 - x + 8 + 8x^3 + 2x - 10$$

Step 2. Rearrange the terms so that like terms are together. (You might do this step mentally.)

$$= 4x^3 + 8x^3 + 3x^2 - x + 2x + 8 - 10$$

Step 3. Systematically combine matching like terms and indicate addition or subtraction of unlike terms.

$$= 12x^3 + 3x^2 + x - 2$$

Step 4. Review the main results.

$$(4x^3 + 3x^2 - x + 8) + (8x^3 + 2x - 10) = 4x^3 + 3x^2 - x + 8 + 8x^3 + 2x - 10$$
$$= 12x^3 + 3x^2 + x - 2$$

Subtraction of Polynomials

To subtract two polynomials, add the opposite of the second polynomial.

You can accomplish subtraction of polynomials by enclosing both polynomials in parentheses and then placing a minus symbol between them. Of course, make sure that the minus symbol precedes the polynomial that is being subtracted.

Problem Perform the indicated subtraction.

a. $(9x^2 - 6x + 2) - (-7x^2 - 5x + 3)$

b. $(4x^3 + 3x^2 - x + 8) - (8x^3 + 2x - 10)$

Solution

a. $(9x^2 - 6x + 2) - (-7x^2 - 5x + 3)$

Be careful with signs! Sign errors are common mistakes for beginning algebra students.

Step 1. Remove parentheses.

$$(9x^2 - 6x + 2) - (-7x^2 - 5x + 3)$$
$$= 9x^2 - 6x + 2 + 7x^2 + 5x - 3$$

Step 2. Systematically combine matching like terms and indicate addition or subtraction of unlike terms.

$$= 16x^2 - x - 1$$

Step 3. Review the main results.

$$(9x^2 - 6x + 2) - (-7x^2 - 5x + 3)$$
$$= 9x^2 - 6x + 2 + 7x^2 + 5x - 3 = 16x^2 - x - 1$$

b. $(4x^3 + 3x^2 - x + 8) - (8x^3 + 2x - 10)$

Step 1. Remove parentheses.

$$(4x^3 + 3x^2 - x + 8) - (8x^3 + 2x - 10)$$
$$= 4x^3 + 3x^2 - x + 8 - 8x^3 - 2x + 10$$

Step 2. Systematically combine matching like terms and indicate addition or subtraction of unlike terms.

$$= -4x^3 + 3x^2 - 3x + 18$$

Step 3. Review the main results.

$$\left(4x^3 + 3x^2 - x + 8\right) - \left(8x^3 + 2x - 10\right) = 4x^3 + 3x^2 - x + 8 - 8x^3 - 2x + 10$$
$$= -4x^3 + 3x^2 - 3x + 18$$

Exercise 8

For 1–5, state the most specific name for the given polynomial.

1. $x^2 - x + 1$
2. $125x^3 - 64y^3$
3. $2x^2 + 7x - 4$
4. $-\frac{1}{3}x^5y^2$
5. $2x^4 + 3x^3 - 7x^2 - x + 8$

For 6–14, simplify.

6. $-15x + 17x$
7. $14xy^3 - 7x^3y^2$
8. $10x^2 - 2x^2 - 20x^2$
9. $10 + 10x$
10. $12x^3 - 5x^2 + 10x - 60 + 3x^3 - 7x^2 - 1$
11. $(10x^2 - 5x + 3) + (6x^2 + 5x - 13)$
12. $(20x^3 - 3x^2 - 2x + 5) + (9x^3 + x^2 + 2x - 15)$
13. $(10x^2 - 5x + 3) - (6x^2 + 5x - 13)$
14. $(20x^3 - 3x^2 - 2x + 5) - (9x^3 + x^2 + 2x - 15)$

9

Multiplying Polynomials

This chapter presents rules for multiplying polynomials. You use the properties of real numbers and the rules of exponents when you multiply polynomials.

Multiplying Monomials

Multiplying Monomials

To multiply monomials, (1) multiply the numerical coefficients, (2) multiply the variable factors using rules for exponents, and (3) use the product of the numerical coefficients as the coefficient of the product of the variable factors to obtain the answer.

Problem Find the product.

a. $(5x^5y^3)(3x^2y^6)$

b. $(-2a^3b^4)(8ab^2)$

c. $(6x)(-2x)$

d. $(-10x^3)(4x^2)$

e. $(4x^2y^5)(-2xy^3)(-3xy)$

f. $(x)(2)$

Solution

a. $(5x^5y^3)(3x^2y^6)$

Step 1. Multiply the numerical coefficients.

$$(5)(3)=15$$

Step 2. Multiply the variable factors.

$$(x^5y^3)(x^2y^6)=x^7y^9$$

Step 3. Use the product in step 1 as the coefficient of x^7y^9.

$$(5x^5y^3)(3x^2y^6)=15x^7y^9$$

To streamline your work when you are multiplying polynomials, arrange the variables in each term alphabetically.

Recall from Chapter 7 that when you multiply exponential expressions that have the same base, you *add* the exponents.

b. $(-2a^3b^4)(8ab^2)$

Step 1. Multiply the numerical coefficients.

$$(-2)(8)=-16$$

Step 2. Multiply the variable factors.

$$(a^3b^4)(ab^2)=a^4b^6$$

If no exponent is written on a variable, the exponent is understood to be 1.

Step 3. Use the product in step 1 as the coefficient of a^4b^6.

$$(-2a^3b^4)(8ab^2)=-16a^4b^6$$

c. $(6x)(-2x)$

Step 1. Multiply the numerical coefficients.

$$(6)(-2)=-12$$

Step 2. Multiply the variable factors.

$$(x)(x)=x^2$$

Step 3. Use the product in step 1 as the coefficient of x^2.

$$(6x)(-2x)=-12x^2$$

d. $(-10x^3)(4x^2)$

Step 1. Multiply the numerical coefficients.

$$(-10)(4)=-40$$

Step 2. Multiply the variable factors.

$$\left(x^3\right)\left(x^2\right)=x^5$$

Step 3. Use the product in step 1 as the coefficient of x^5.

$$\left(-10x^3\right)\left(4x^2\right)=-40x^5$$

e. $(4x^2y^5)(-2xy^3)(-3xy)$

Step 1. Multiply the numerical coefficients.

$$(4)(-2)(-3)=24$$

Step 2. Multiply the variable factors.

$$\left(x^2y^5\right)\left(xy^3\right)\left(xy\right)=x^4y^9$$

Step 3. Use the product in step 1 as the coefficient of x^4y^9.

$$\left(4x^2y^5\right)\left(-2xy^3\right)\left(-3xy\right)=24x^4y^9$$

f. $(x)(2)$

Step 1. Multiply the numerical coefficients.

$$(1)(2)=2$$

Step 2. Multiply the variable factors.

There is only one variable factor, x.

Step 3. Use the product in step 1 as the coefficient of x.

$$(x)(2)=2x$$

Multiplying Polynomials by Monomials

Multiplying a Polynomial by a Monomial

To multiply a polynomial by a monomial, multiply each term of the polynomial by the monomial.

This rule is a direct application of the distributive property for real numbers.

Problem Find the product.

a. $2(x+5)$

b. $x(3x-2)$

c. $-8a^3b^5(2a^2-7ab^2-3)$

d. $x^2(2x^4+4x^3-3x+6)$

Solution

a. $2(x+5)$

Step 1. Multiply each term of the polynomial by the monomial.

$$2(x+5)$$
$$=2\cdot x+2\cdot 5$$
$$=2x+10$$

b. $x(3x-2)$

Step 1. Multiply each term of the polynomial by the monomial.

$$x(3x-2)$$
$$=x\cdot 3x-x\cdot 2$$
$$=3x^2-2x$$

c. $-8a^3b^5(2a^2-7ab^2-3)$

Step 1. Multiply each term of the polynomial by the monomial.

Be careful! Watch your exponents when you are multiplying polynomials by monomials.

$$-8a^3b^5\left(2a^2-7ab^2-3\right)$$
$$=\left(-8a^3b^5\right)\left(2a^2\right)-\left(-8a^3b^5\right)\left(7ab^2\right)-\left(-8a^3b^5\right)(3)$$
$$=-16a^5b^5+56a^4b^7+24a^3b^5$$

d. $x^2(2x^4+4x^3-3x+6)$

Step 1. Multiply each term of the polynomial by the monomial.

$$x^2(2x^4+4x^3-3x+6)$$
$$=x^2\cdot 2x^4+x^2\cdot 4x^3-x^2\cdot 3x+x^2\cdot 6$$
$$=2x^6+4x^5-3x^3+6x^2$$

Multiplying Binomials

Multiplying Two Binomials

To multiply two binomials, multiply all the terms of the second binomial by each term of the first binomial and then simplify.

Problem Find the product.

a. $(2x+3)(x-5)$

b. $(a+b)(c+d)$

Solution

a. $(2x+3)(x-5)$

Step 1. Multiply all the terms of the second binomial by each term of the first binomial.

$$(2x+3)(x-5)$$

$$= 2x\cdot x-2x\cdot 5+3\cdot x-3\cdot 5$$

$$= 2x^2-10x+3x-15$$

$(2x+3)(x-5)\neq 2x^2-15$. Don't forget about $-2x\cdot 5+3\cdot x$!

Step 2. Simplify.

$$=2x^2-7x-15$$

b. $(a+b)(c+d)$

Step 1. Multiply all the terms of the second binomial by each term of the first binomial.

$$(a+b)(c+d)$$

$$=a\cdot c+a\cdot d+b\cdot c+b\cdot d$$

$$=ac+ad+bc+bd$$

Step 2. Simplify.

There are no like terms, so $ac+ad+bc+bd$ is simplified.

The FOIL Method

From the last problem, you can see that to find the product of two binomials, you compute four products, called *partial products*, using the terms of the two binomials. The FOIL method is a quick way to get those four partial products. Here is how FOIL works for finding the four partial products for $(a+b)(c+d)$.

1. Multiply the two **F**irst terms: $a \cdot c$.
2. Multiply the two **O**uter terms: $a \cdot d$.
3. Multiply the two **I**nner terms: $b \cdot c$.
4. Multiply the two **L**ast terms: $b \cdot d$.

Be aware that the FOIL method works only for the product of two binomials.

Forgetting to compute the middle terms is the most common error when finding the product of two binomials.

Notice that FOIL is an acronym for first, outer, inner, and last. The inner and outer partial products are the *middle terms*.

Problem Find the product using the FOIL method.

a. $(5x+4)(2x-3)$

b. $(x-2)(x-5)$

c. $(x-2)(x+5)$

d. $(x+y)^2$

Solution

a. $(5x+4)(2x-3)$

Step 1. Find the partial products using the acronym FOIL.

$$(5x+4)(2x-3)$$

$$= \underbrace{5x \cdot 2x}_{\text{First}} - \underbrace{5x \cdot 3}_{\text{Outer}} + \underbrace{4 \cdot 2x}_{\text{Inner}} - \underbrace{4 \cdot 3}_{\text{Last}} = 10x^2 \ \underbrace{-15x + 8x}_{\text{Middle terms}} \ -12$$

Step 2. Simplify.

$$= 10x^2 - 7x - 12$$

Step 3. State the main result.

$$(5x+4)(2x-3) = 10x^2 - 7x - 12$$

b. $(x - 2)(x - 5)$

Step 1. Find the partial products using the acronym FOIL.

$$(x-2)(x-5)$$

$$= \underbrace{x \cdot x}_{\text{First}} - \underbrace{x \cdot 5}_{\text{Outer}} - \underbrace{2 \cdot x}_{\text{Inner}} + \underbrace{2 \cdot 5}_{\text{Last}} = x^2 \underbrace{-5x - 2x}_{\text{Middle terms}} + 10$$

Step 2. Simplify.

$$= x^2 - 7x + 10$$

Step 3. State the main result.

$$(x-2)(x-5) = x^2 - 7x + 10$$

c. $(x - 2)(x + 5)$

Step 1. Find the partial products using the acronym FOIL.

$$(x-2)(x+5)$$

$$= \underbrace{x \cdot x}_{\text{First}} + \underbrace{x \cdot 5}_{\text{Outer}} - \underbrace{2 \cdot x}_{\text{Inner}} - \underbrace{2 \cdot 5}_{\text{Last}} = x^2 + \underbrace{5x - 2x}_{\text{Middle terms}} - 10$$

Step 2. Simplify.

$$= x^2 + 3x - 10$$

Step 3. State the main result.

$$(x-2)(x+5) = x^2 + 3x - 10$$

d. $(x + y)^2$

Step 1. Write as a product.

$$(x+y)^2 = (x+y)(x+y)$$

Step 2. Find the partial products using the acronym FOIL.

$$= \underbrace{x \cdot x}_{\text{First}} + \underbrace{x \cdot y}_{\text{Outer}} + \underbrace{y \cdot x}_{\text{Inner}} + \underbrace{y \cdot y}_{\text{Last}} = x^2 + \underbrace{xy + xy}_{\text{Middle terms}} + y^2$$

Step 3. Simplify.

$$= x^2 + 2xy + y^2$$

> $(x + y)^2 \neq x^2 + y^2$! Don't forget the middle terms!

Step 4. State the main result.

$$(x+y)^2 = (x+y)(x+y) = x^2 + 2xy + y^2$$

Multiplying Polynomials

Multiplying Two Polynomials

To multiply two polynomials, multiply all the terms of the second polynomial by each term of the first polynomial and then simplify.

Problem Find the product.

a. $(2x - 1)(3x^2 - 5x + 4)$

b. $(4x^2 + 2x - 5)(2x^2 - x - 3)$

c. $(x - 2)(x^2 + 2x + 4)$

Solution

a. $(2x - 1)(3x^2 - 5x + 4)$

Step 1. Multiply all the terms of the second polynomial by each term of the first polynomial.

$$(2x-1)(3x^2 - 5x + 4) = 2x \cdot 3x^2 - 2x \cdot 5x + 2x \cdot 4 - 1 \cdot 3x^2 + 1 \cdot 5x - 1 \cdot 4$$

$$= 6x^3 - 10x^2 + 8x - 3x^2 + 5x - 4$$

Step 2. Simplify.

$$= 6x^3 - 13x^2 + 13x - 4$$

b. $(4x^2 + 2x - 5)(2x^2 - x - 3)$

Step 1. Multiply all the terms of the second polynomial by each term of the first polynomial.

$$(4x^2 + 2x - 5)(2x^2 - x - 3)$$

$$= 4x^2 \cdot 2x^2 - 4x^2 \cdot x - 4x^2 \cdot 3 + 2x \cdot 2x^2 - 2x \cdot x - 2x \cdot 3 - 5 \cdot 2x^2 + 5 \cdot x + 5 \cdot 3$$

$$= 8x^4 - 4x^3 - 12x^2 + 4x^3 - 2x^2 - 6x - 10x^2 + 5x + 15$$

Step 2. Simplify.

$$= 8x^4 - 24x^2 - x + 15$$

c. $(x-2)(x^2+2x+4)$

Step 1. Multiply all the terms of the second polynomial by each term of the first polynomial.

$$(x-2)(x^2+2x+4)$$

$$= x \cdot x^2 + x \cdot 2x + x \cdot 4 - 2 \cdot x^2 - 2 \cdot 2x - 2 \cdot 4$$

$$= x^3 + 2x^2 + 4x - 2x^2 - 4x - 8$$

Step 2. Simplify.

$$= x^3 - 8$$

Special Products

The answer to the last problem is an example of the "difference of two cubes." It is a special product. Here is a list of *special products* that you need to know for algebra.

Special Products

Memorizing special products is a winning strategy in algebra.

Perfect Squares

$(x+y)^2 = x^2 + 2xy + y^2$

$(x-y)^2 = x^2 - 2xy + y^2$

Difference of Two Squares

$(x+y)(x-y) = x^2 - y^2$

Perfect Cubes

$(x+y)^3 = x^3 + 3x^2y + 3xy^2 + y^3$

$(x-y)^3 = x^3 - 3x^2y + 3xy^2 - y^3$

Sum of Two Cubes

$(x+y)(x^2 - xy + y^2) = x^3 + y^3$

Difference of Two Cubes

$(x-y)(x^2 + xy + y^2) = x^3 - y^3$

Exercise 9

Find the product.

1. $(4x^5y^3)(-3x^2y^3)$
2. $(-8a^4b^3)(5ab^2)$
3. $(-10x^3)(-2x^2)$
4. $(-3x^2y^5)(6xy^4)(-2xy)$
5. $3(x-5)$
6. $x(3x^2-4)$
7. $-2a^2b^3(3a^2-5ab^2-10)$
8. $(2x-3)(x+4)$
9. $(x+4)(x+5)$
10. $(x-4)(x-5)$
11. $(x+4)(x-5)$
12. $(x-4)(x+5)$
13. $(x-1)(2x^2-5x+3)$
14. $(2x^2+x-3)(5x^2-x-2)$
15. $(x-y)^2$
16. $(x+y)(x-y)$
17. $(x+y)^3$
18. $(x-y)^3$
19. $(x+y)(x^2-xy+y^2)$
20. $(x-y)(x^2+xy+y^2)$

10

Simplifying Polynomial Expressions

In this chapter, you apply your skills in multiplying polynomials to the process of simplifying polynomial expressions.

Identifying Polynomials

A *polynomial expression* is composed of polynomials only and can contain grouping symbols, multiplication, addition, subtraction, and raising to nonzero powers only.

No division by polynomials or raising polynomials to negative powers is allowed in a polynomial expression.

Problem Specify whether the expression is a polynomial expression.

a. $5+2(a-5)$

b. $-8xy^3+\dfrac{5}{2x^2}-27$

c. $(2x-1)(3x-4)+(x-1)^2$

d. $4(x^{-3}-y^2)+5(x+y^{-1})$

e. $\dfrac{x^2+y^2}{x^2-y^2}$

f. $2x^2-x-4[3x+5(x-4)]$

Solution

a. $5 + 2(a - 5)$

Step 1. Check whether the expression meets the criteria for a polynomial expression.

$5+2(a-5)$ is composed of polynomials and contains permissible components, so it is a polynomial expression.

b. $-8xy^3 + \dfrac{5}{2x^2} - 27$

Step 1. Check whether the expression meets the criteria for a polynomial expression.

$-8xy^3 + \dfrac{5}{2x^2} - 27$ is not a polynomial expression because it contains division by $2x^2$.

c. $(2x - 1)(3x - 4) + (x - 1)^2$

Step 1. Check whether the expression meets the criteria for a polynomial expression.

$(2x-1)(3x-4)+(x-1)^2$ is composed of polynomials and contains permissible components, so it is a polynomial expression.

d. $4(x^{-3} - y^2) + 5(x + y^{-1})$

Step 1. Check whether the expression meets the criteria for a polynomial expression.

$4\left(x^{-3}-y^2\right)+5\left(x+y^{-1}\right)$ is not a polynomial expression because it is not composed of polynomials.

e. $\dfrac{x^2 + y^2}{x^2 - y^2}$

Step 1. Check whether the expression meets the criteria for a polynomial expression.

$\dfrac{x^2+y^2}{x^2-y^2}$ is not a polynomial expression because it contains division by a polynomial.

f. $2x^2 - x - 4[3x + 5(x - 4)]$

Step 1. Check whether the expression meets the criteria for a polynomial expression.

$2x^2 - x - 4[3x + 5(x - 4)]$ is composed of polynomials and contains permissible components, so it is a polynomial expression.

Simplifying Polynomials

When you simplify polynomial expressions, you proceed in an orderly fashion so that you do not violate the order of operations for real numbers. After all, the variables in polynomials are simply stand-ins for real numbers, so it is important that what you do is consistent with the rules for working with real numbers.

Simplifying Polynomial Expressions

To simplify a polynomial expression:

1. Simplify within grouping symbols, if any. Start with the innermost grouping symbol and work outward.
2. Do powers, if indicated.
3. Do multiplication, if indicated.
4. Simplify the result.

Problem Simplify.

a. $5 + 2(a - 5)$

b. $-3(y + 4) + 8y$

c. $9xy - x(3y - 5x) - 2x^2$

d. $(2x - 1)(3x - 4) + (x - 1)^2$

e. $2x^2 - x - 4[3x + 5(x - 4)]$

f. $2(x + 1)^2$

Solution

a. $5 + 2(a - 5)$

Step 1. Do multiplication: $2(a - 5)$.

$$5 + 2(a - 5)$$
$$= 5 + \mathbf{2a - 10}$$

Step 2. Simplify the result.

$= \mathbf{2a - 5}$

Step 3. Review the main steps.

$5 + 2(a-5) = 5 + 2a - 10 = 2a - 5$

5 + 2(a − 5) ≠ 7(a − 5). Do multiplication before addition, if no parentheses indicate otherwise.

b. $-3(y+4)+8y$

Step 1. Do multiplication: $-3(y+4)$.

$-3(y+4)+8y$

$= \mathbf{-3y - 12} + 8y$

Step 2. Simplify the result.

$= \mathbf{5y} - 12$

Step 3. Review the main steps.

$-3(y+4)+8y = -3y - 12 + 8y = 5y - 12$

c. $9xy - x(3y-5x) - 2x^2$

Step 1. Do multiplication: $-x(3y-5x)$.

$9xy - x(3y-5x) - 2x^2$

$= 9xy - \mathbf{3xy} + \mathbf{5x^2} - 2x^2$

Step 2. Simplify the result.

$= \mathbf{3x^2 + 6xy}$

Step 3. Review the main steps.

$9xy - x(3y-5x) - 2x^2 = 9xy - 3xy + 5x^2 - 2x^2 = 3x^2 + 6xy$

d. $(2x-1)(3x-4)+(x-1)^2$

Step 1. Do the power: $(x-1)^2$.

$(2x-1)(3x-4)+(x-1)^2$

$= (2x-1)(3x-4) + \mathbf{x^2} - \mathbf{2x} + \mathbf{1}$

Step 2. Do multiplication: $(2x-1)(3x-4)$.

$$=\mathbf{6x^2-11x+4}+x^2-2x+1$$

Step 3. Simplify the results.

$$=\mathbf{7x^2-13x+5}$$

Step 4. Review the main steps.

$$(2x-1)(3x-4)+(x-1)^2=6x^2-11x+4+x^2-2x+1=7x^2-13x+5$$

e. $2x^2-x-4[3x+5(x-4)]$

Step 1. Simplify within the brackets. First, do multiplication: $5(x-4)$.

$$2x^2-x-4[3x+5(x-4)]$$
$$=2x^2-x-4[3x+\mathbf{5x-20}]$$

Step 2. Simplify $3x+5x-20$ within the brackets.

$$=2x^2-x-4[\mathbf{8x-20}]$$

Step 3. Do multiplication: $-4[8x-20]$

$$=2x^2-x\mathbf{-32x+80}$$

Step 4. Simplify the result.

$$=2x^2\mathbf{-33x}+80$$

Step 5. Review the main steps.

$$2x^2-x-4[3x+5(x-4)]=2x^2-x-4[3x+5x-20]$$
$$=2x^2-x-4[8x-20]$$
$$=2x^2-x-32x+80=2x^2-33x+80$$

f. $2(x+1)^2$

Step 1. Do the power: $(x+1)^2$.

$$=2(\mathbf{x^2+2x+1})$$

$2(x+1)^2 \neq (2x+2)^2$. The exponent applies only to $(x+1)$.

Step 2. Do multiplication: $2\left(x^2+2x+1\right)$

$$=\mathbf{2x^2+4x+2}$$

Step 3. Review the main steps.

$$2(x+1)^2=2\left(x^2+2x+1\right)=2x^2+4x+2$$

Exercise 10

Simplify.

1. $8+2(x-5)$
2. $-7(y-4)+9y$
3. $10xy-x(5y-3x)-4x^2$
4. $(3x-1)(2x-5)+(x+1)^2$
5. $3x^2-4x-5[x-2(x-8)]$
6. $-x(x+4)+5(x-2)$
7. $(a-5)(a+2)-(a-6)(a-4)$
8. $5x^2-(-3xy-2y^2)$
9. $x^2-[2x-x(3x-1)]+6x$
10. $(4x^2y^5)(-2xy^3)(-3xy)-15x^2y^3(2x^2y^6+2)$

11

Dividing Polynomials

This chapter presents a discussion of division of polynomials. Division of polynomials is analogous to division of real numbers. In algebra, you indicate division using the fraction bar. For example, $\frac{16x^3 - 28x^2}{-4x}$, $x \neq 0$, indicates $16x^3 - 28x^2$ divided by $-4x$. Because division by 0 is undefined, you must exclude values for the variable or variables that would make the divisor 0. For convenience, you can assume such values are excluded as you work through the problems in this chapter.

Dividing a Polynomial by a Monomial

Customarily, a division problem is a *dividend* divided by a *divisor*. When you do the division, you get a *quotient* and a *remainder*. You express the relationship between these quantities as

$$\frac{\text{dividend}}{\text{divisor}} = \text{quotient} + \frac{\text{remainder}}{\text{divisor}}.$$

Be sure to note that the remainder is the *numerator* of the expression $\frac{\text{remainder}}{\text{divisor}}$.

Dividing a Polynomial by a Monomial

To divide a polynomial by a monomial, divide each term of the polynomial by the monomial.

To avoid sign errors when you are doing division of polynomials, *keep a – symbol with the number that follows it.* You likely will need to properly insert a + symbol when you do this.

Sign errors are a major reason for mistakes in division of polynomials.

You will see this tactic illustrated in the following problem.

Problem Find the quotient and remainder.

a. $\dfrac{16x^3 - 28x^2}{-4x}$

b. $\dfrac{-12x^4 + 6x^2}{-3x}$

c. $\dfrac{4x^4y - 8x^3y^3 + 16xy^4}{4xy}$

d. $\dfrac{6x^4 + 1}{2x^4}$

e. $\dfrac{16x^5y^2}{16x^5y^2}$

Solution

a. $\dfrac{16x^3 - 28x^2}{-4x}$

Step 1. Divide each term of the polynomial by the monomial.

$$\frac{16x^3 - 28x^2}{-4x}$$

Keep with 28 ↓

$$= \frac{16x^3}{-4x} + \frac{-28x^2}{-4x}$$

↑ Insert

$$= -4x^2 + 7x$$

Step 2. State the quotient and remainder.

The quotient is $-4x^2 + 7x$ and the remainder is 0.

b. $\dfrac{-12x^4 + 6x^2}{-3x}$

Step 1. Divide each term of the polynomial by the monomial.

$$\frac{-12x^4 + 6x^2}{-3x}$$

$$= \frac{-12x^4}{-3x} + \frac{6x^2}{-3x}$$

$$= 4x^3 - 2x$$

Step 2. State the quotient and remainder.

The quotient is $4x^3 - 2x$ and the remainder is 0.

c. $\dfrac{4x^4y - 8x^3y^3 + 16xy^4}{4xy}$

Step 1. Divide each term of the polynomial by the monomial.

$$\frac{4x^4y - 8x^3y^3 + 16xy^4}{4xy}$$

$$= \frac{4x^4y}{4xy} + \frac{-8x^3y^3}{4xy} + \frac{16xy^4}{4xy}$$

$$= x^3 - 2x^2y^2 + 4y^3$$

Step 2. State the quotient and remainder.

The quotient is $x^3 - 2x^2y^2 + 4y^3$ and the remainder is 0.

d. $\dfrac{6x^4 + 1}{2x^4}$

Step 1. Divide each term of the polynomial by the monomial.

$$\frac{6x^4 + 1}{2x^4}$$

$$= \frac{6x^4}{2x^4} + \frac{1}{2x^4}$$

$$= 3 + \frac{1}{2x^4} \begin{array}{l} \leftarrow \text{remainder} \\ \leftarrow \text{divisor} \end{array}$$

$\uparrow$ quotient

Step 2. State the quotient and remainder.

The quotient is 3 and the remainder is 1.

e. $\dfrac{16x^5y^2}{16x^5y^2}$

Step 1. Divide $16x^5y^2$ by $16x^5y^2$.

$$\frac{16x^5y^2}{16x^5y^2} = 1$$

Step 2. State the quotient and remainder.

The quotient is 1 and the remainder is 0.

Dividing a Polynomial by a Polynomial

When you divide two polynomials, and the divisor is not a monomial, you use long division. The procedure is very similar to the long division algorithm of arithmetic. The steps are illustrated in the following problem.

Problem Find the quotient and remainder.

a. $\dfrac{4x^3 + 8x - 6x^2 + 1}{2x - 1}$

b. $\dfrac{x^3 - 8}{x - 2}$

c. $\dfrac{x^2 + x - 4}{x + 3}$

Solution

a. $\dfrac{4x^3 + 8x - 6x^2 + 1}{2x - 1}$

Step 1. Using the long division symbol $(\overline{)}\ )$, arrange the terms of both the dividend and the divisor in descending powers of the variable x.

$$\frac{4x^3 + 8x - 6x^2 + 1}{2x - 1}$$

$$= 2x - 1\overline{)4x^3 - 6x^2 + 8x + 1}$$

Step 2. Divide the first term of the dividend by the first term of the divisor and write the answer as the first term of the quotient.

$$\begin{array}{r} \mathbf{2x^2} \\ 2x - 1\overline{)4x^3 - 6x^2 + 8x + 1} \end{array}$$

Step 3. Multiply $2x - 1$ by $2x^2$ and enter the product under the dividend.

$$\begin{array}{l} 2x^2 \\ 2x - 1\overline{)4x^3 - 6x^2 + 8x + 1} \\ \mathbf{4x^3 - 2x^2} \end{array}$$

Step 4. Subtract $4x^3 - 2x^2$ from the dividend, being sure to mentally change the signs of *both* $4x^3$ and $-2x^2$.

In long division of polynomials, making sign errors when subtracting is the most common mistake.

$$\begin{array}{l} \mathbf{2x^2} \\ 2x - 1\overline{)4x^3 - 6x^2 + 8x + 1} \\ \underline{4x^3 - 2x^2} \\ \mathbf{-4x^2} \end{array}$$

Step 5. Bring down $8x$, the next term of the dividend, and repeat steps 2–4.

$$\begin{array}{l} 2x^2 - \mathbf{2x} \\ 2x - 1\overline{)4x^3 - 6x^2 + 8x + 1} \\ \underline{4x^3 - 2x^2} \\ -4x^2 + \mathbf{8x} \\ \underline{\mathbf{-4x^2 + 2x}} \\ \mathbf{6x} \end{array}$$

Step 6. Bring down 1, the last term of the dividend, and repeat steps 2–4.

$$
\begin{array}{r}
2x^2 - 2x + \mathbf{3} \\
2x - 1 \overline{\smash{)}\, 4x^3 - 6x^2 + 8x + 1} \\
\underline{4x^3 - 2x^2} \quad\quad\quad \\
-4x^2 + 8x \quad \\
\underline{-4x^2 + 2x} \quad \\
6x + \mathbf{1} \\
\underline{\mathbf{6x - 3}} \\
\mathbf{4}
\end{array}
$$

Step 7. State the quotient and remainder.

The quotient is $2x^2 - 2x + 3$ and the remainder is 4.

b. $\dfrac{x^3 - 8}{x - 2}$

Step 1. Using the long division symbol $\left(\overline{)}\;\right)$, arrange the terms of both the dividend and the divisor in descending powers of the variable x. Insert zeros as placeholders for missing powers of x.

$$\frac{x^3 - 8}{x - 2}$$

$$= x - 2 \overline{\smash{)}\, x^3 + 0 + 0 - 8}$$

$\dfrac{x^3 - 8}{x - 2} \neq x^2 + 4$. Avoid this common error.

Step 2. Divide the first term of the dividend by the first term of the divisor and write the answer as the first term of the quotient.

$$
\begin{array}{r}
\mathbf{x^2} \quad\quad\quad\quad \\
x - 2 \overline{\smash{)}\, x^3 + 0 + 0 - 8}
\end{array}
$$

Step 3. Multiply $x - 2$ by x^2 and enter the product under the dividend.

$$
\begin{array}{r}
x^2 \quad\quad\quad\quad \\
x - 2 \overline{\smash{)}\, x^3 + 0 + 0 - 8} \\
\mathbf{x^3 - 2x^2} \quad\quad
\end{array}
$$

Step 4. Subtract $x^3 - 2x^2$ from the dividend, being sure to mentally change the signs of *both* x^3 and $-2x^2$.

$$\begin{array}{r} x^2 \\ x-2\overline{)x^3+0+0-8} \\ \underline{x^3-2x^2} \\ \mathbf{2x^2} \end{array}$$

Step 5. Bring down 0, the next term of the dividend, and repeat steps 2–4.

$$\begin{array}{r} x^2+\mathbf{2x} \\ x-2\overline{)x^3+0+0-8} \\ \underline{x^3-2x^2} \\ 2x^2+\mathbf{0} \\ \underline{\mathbf{2x^2-4x}} \\ \mathbf{4x} \end{array}$$

Step 6. Bring down –8, the last term of the dividend, and repeat steps 2–4.

$$\begin{array}{r} x^2+2x+\mathbf{4} \\ x-2\overline{)x^3+0+0-8} \\ \underline{x^3-2x^2} \\ 2x^2+0 \\ \underline{2x^2-4x} \\ 4x-\mathbf{8} \\ \underline{\mathbf{4x-8}} \\ \mathbf{0} \end{array}$$

Step 7. State the quotient and remainder.

The quotient is $x^2 + 2x + 4$ and the remainder is 0.

c. $\dfrac{x^2 + x - 4}{x + 3}$

Step 1. Using the long division symbol $\left(\overline{)\ }\right)$, arrange the terms of both the dividend and the divisor in descending powers of the variable x.

$$\frac{x^2+x-4}{x+3}$$
$$= x+3\overline{)x^2+x-4}$$

Step 2. Divide the first term of the dividend by the first term of the divisor and write the answer as the first term of the quotient.

$$\begin{array}{r} \boldsymbol{x} \\ x+3\overline{)x^2+x-4} \end{array}$$

Step 3. Multiply $x+3$ by x and enter the product under the dividend.

$$\begin{array}{r} x \\ x+3\overline{)x^2+x-4} \\ \boldsymbol{x^2+3x} \end{array}$$

Step 4. Subtract $x^2 + 3x$ from the dividend, being sure to mentally change the signs of *both* x^2 and $3x$.

$$\begin{array}{r} x \\ x+3\overline{)x^2+x-4} \\ \underline{x^2+3x} \\ \boldsymbol{-2x} \end{array}$$

Step 5. Bring down -4, the next term of the dividend, and repeat steps 2–4.

$$\begin{array}{r} x-\boldsymbol{2} \\ x+3\overline{)x^2+x-4} \\ \underline{x^2+3x} \\ -2x-\boldsymbol{4} \\ \underline{\boldsymbol{-2x-6}} \\ \boldsymbol{2} \end{array}$$

Step 6. State the quotient and remainder.

The quotient is $x-2$ and the remainder is 2.

Exercise 11

Find the quotient and remainder.

1. $\dfrac{15x^5 - 30x^2}{-5x}$

2. $\dfrac{-14x^4 + 21x^2}{-7x^2}$

3. $\dfrac{25x^4y^2}{-5x}$

4. $\dfrac{6x^5y^2 - 8x^3y^3 + 10xy^6}{2xy^2}$

5. $\dfrac{-10x^4y^4z^4 - 20x^2y^5z^2}{10x^2y^3z}$

6. $\dfrac{-18x^5 + 5}{3x^5}$

7. $\dfrac{7a^6b^3 - 14a^5b^2 - 42a^4b^2 + 7a^3b^2}{7a^3b^2}$

8. $\dfrac{x^2 - 1}{x + 1}$

9. $\dfrac{x^2 - 9x + 20}{x - 4}$

10. $\dfrac{2x^3 - 13x + x^2 + 6}{x - 4}$

12

Factoring Polynomials

In this chapter, you learn about factoring polynomials.

Factoring and Its Objectives

Factoring is the process of undoing multiplication, so you need a strong understanding of multiplication of polynomials to be skillful in factoring them. You can expect that, from time to time throughout the chapter, you will be asked to recall your prior knowledge of multiplication of polynomials.

The reality that factoring polynomials requires agility in multiplying polynomials underscores the importance of mastering previous skills in mathematics before going on to new ones. Expecting to learn a new math topic that relies on previous skills that were not mastered is a self-defeating strategy that often leads to disappointing results. Some advice: Don't make this mistake!

The objective in factoring is to take a complicated polynomial and express it as the product of simpler polynomial factors. You might ask, "Why would you want to do this?" One practical answer is that, generally, you have fewer restrictions on factors than you do on terms. Factors are joined by multiplication, but terms are joined by addition or subtraction. The following problem illustrates this point.

Problem Follow the indicated directions for the six true sentences given below. (For convenience, assume x and y are positive real numbers with $x < y$.)

1. $\sqrt{xy} = \sqrt{x} \cdot \sqrt{y}$, but $\sqrt{x+y} \neq \sqrt{x} + \sqrt{y}$.

2. $(xy)^2 = x^2y^2$, but $(x + y)^2 \neq x^2 + y^2$.

3. $|xy| = |x|\,|y|$, but $|x - y| \neq |x| - |y|$.

4. $\frac{xy}{x} = y$, but $\frac{x+y}{x} \neq 1+y$; also, $\frac{x+y}{x} \neq \frac{\cancel{x}+y}{\cancel{x}} \neq y$.

5. $\frac{1}{xy} = \frac{1}{x} \cdot \frac{1}{y}$, but $\frac{1}{x+y} \neq \frac{1}{x} + \frac{1}{y}$.

6. $\frac{1}{x^{-2}y^{-2}} = x^2y^2$, but $\frac{1}{x^{-2}+y^{-2}} \neq x^2 + y^2$.

a. From the statements of equality and inequality in sentences 1–6, list those involving terms.

b. From the statements of equality and inequality in sentences 1–6, list those involving factors and no terms.

Solution

a. From the statements of equality and inequality in sentences 1–6, list those involving terms.

Step 1. Examine sentences 1–6 for equality or inequality statements involving terms and then list those involving terms.

$\sqrt{x+y} \neq \sqrt{x} + \sqrt{y}$, $(x+y)^2 \neq x^2 + y^2$, $|x-y| \neq |x| - |y|$, $\frac{x+y}{x} \neq 1+y$, $\frac{1}{x+y} \neq \frac{1}{x} + \frac{1}{y}$, $\frac{1}{x^{-2}+y^{-2}} \neq x^2 + y^2$ (Note that the statements involving terms contain the $\neq$ symbol.)

b. From the statements of equality and inequality in sentences 1–6, list those involving factors and no terms.

Step 1. Examine sentences 1–6 for equality or inequality statements involving terms and then list those not involving terms.

$\sqrt{xy} = \sqrt{x} \cdot \sqrt{y}$, $(xy)^2 = x^2y^2$, $|xy| = |x||y|$, $\frac{xy}{x} = y$, $\frac{1}{xy} = \frac{1}{x} \cdot \frac{1}{y}$, $\frac{1}{x^{-2}y^{-2}} = x^2y^2$ (Note that the statements involving factors and no terms contain the $=$ symbol.)

Greatest Common Factor

The previous problem should motivate you to become proficient in factoring polynomials. The discussion begins with factoring out the greatest common monomial factor. The *greatest common monomial factor* is the product of the greatest common numerical factor and a second component made up of the common variable factors, each with the highest power common to each term. You can refer to the greatest common monomial factor as the *greatest common factor* (GCF).

Problem Find the GCF for the terms in the polynomial $12x^8y^3 - 8x^6y^7z^2$.

Solution

Step 1. Find the numerical factor of the GCF by finding the greatest common numerical factor of 12 and 8.

The factors of 12 are 1, 2, 3, **4**, 6, and 12, and the factors of 8 are 1, 2, **4**, and 8. The numerical factor of the GCF is 4.

Step 2. Identify the common variable factors, each with the highest power common to x^8y^3 and $x^6y^7z^2$.

x and y are the common variable factors. The highest power of x that is common to each term is x^6, and the highest power of y that is common to each term is y^3. The common variable component of the GCF is x^6y^3.

Step 3. Write the GCF as the product of the results of steps 1 and 2.

The GCF for the terms in the polynomial $12x^8y^3 - 8x^6y^7z^2$ is $4x^6y^3$.

Problem Factor.

When factoring out the GCF, check your work by mentally multiplying the factors of your answers.

a. $12x^8y^3 - 8x^6y^7z^2$

b. $15x^2 - 3x$

c. $x^3y - xy + y$

d. $4x + 4y$

Solution

a. $12x^8y^3 - 8x^6y^7z^2$

Step 1. Determine the GCF for $12x^8y^3$ and $8x^6y^7z^2$.

GCF $= 4x^6y^3$

Step 2. Rewrite each term of the polynomial as an equivalent product of $4x^6y^3$ and a second factor.

$12x^8y^3 - 8x^6y^7z^2$

$$= \mathbf{4x^6y^3} \cdot 3x^2 - \mathbf{4x^6y^3} \cdot 2y^4z^2$$

Step 3. Use the distributive property to factor $4x^6y^3$ from the resulting expression.

$= \mathbf{4x^6y^3}(3x^2 - 2y^4z^2)$

$12x^8y^3 - 8x^6y^7z^2$ is not in factored form because it is *not* a product, but $4x^6y^3(3x^2 - 2y^4z^2)$ *is* in factored form because it is the product of $4x^6y^3$ and $(3x^2 - 2y^4z^2)$.

Step 3. Review the main steps.

$$12x^8y^3 - 8x^6y^7z^2 = 4x^6y^3\left(3x^2 - 2y^4z^2\right)$$

b. $15x^2 - 3x$

Step 1. Determine the GCF for $15x^2$ and $3x$.

GCF $= 3x$

Step 2. Rewrite each term of the polynomial as an equivalent product of $3x$ and a second factor.

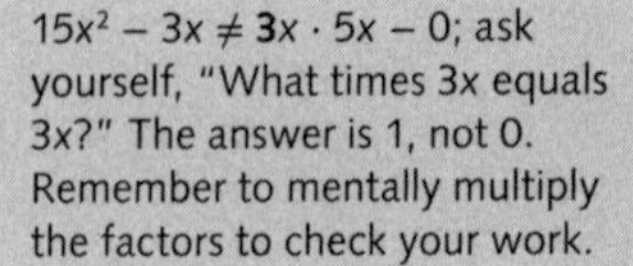
$15x^2 - 3x \neq \mathbf{3x} \cdot 5x - 0$; ask yourself, "What times 3x equals 3x?" The answer is 1, not 0. Remember to mentally multiply the factors to check your work.

$$15x^2 - 3x$$
$$= \mathbf{3x} \cdot 5x - \mathbf{3x} \cdot 1$$

Step 3. Use the distributive property to factor $3x$ from the resulting expression.

$= \mathbf{3x}(5x - 1)$

Step 4. Review the main steps.

$$15x^2 - 3x = 3x(5x - 1)$$

c. $x^3y - xy + y$

Step 1. Determine the GCF for x^3y, xy, and y.

GCF $= y$

Step 2. Rewrite each term of the polynomial as an equivalent product of y and a second factor.

$$x^3y - xy + y$$
$$= \mathbf{y} \cdot x^3 - \mathbf{y} \cdot x + \mathbf{y} \cdot 1$$

Step 3. Use the distributive property to factor y from the resulting expression.

$= \mathbf{y}(x^3 - x + 1)$

Step 4. Review the main steps.

$$x^3y - xy + y = y(x^3 - x + 1)$$

$x^3y - xy + y \neq y(x^3 - x)$. Don't forget the 1.

d. $4x + 4y$

Step 1. Use the distributive property to factor 4 from the polynomial.

$$4x + 4y$$
$$= 4(x + y)$$

GCF with a Negative Coefficient

At times, you might need to factor out a GCF that has a negative coefficient. To avoid sign errors, mentally change subtraction to *add the opposite.*

When factoring out a GCF that has a negative coefficient, *always* mentally multiply the factors and check the signs.

Problem Factor using a negative coefficient for the GCF.

a. $-5xy^2 + 10xy$

b. $-5xy^2 - 10xy$

c. $-x - y$

d. $-2x^3 + 4x - 8$

Solution

a. $-5xy^2 + 10xy$

Step 1. Determine the GCF with a negative coefficient for $-5xy^2$ and $10xy$.

$$\text{GCF} = -5xy$$

Step 2. Rewrite each term of the polynomial as an equivalent product of $-5xy$ and a second factor.

$$-5xy^2 + 10xy$$
$$= \mathbf{-5xy} \cdot y - \mathbf{5xy} \cdot -2$$

$-5xy^2 + 10xy \neq -5xy \cdot y - 5xy \cdot 2$. Check the signs!

Step 3. Use the distributive property to factor $-5xy$ from the resulting expression.

$$= \mathbf{-5xy}(y - 2)$$

Step 4. Review the main steps.

$$-5xy^2 + 10xy = -5xy(y - 2)$$

b. $-5xy^2 - 10xy$

Step 1. Determine the GCF with a negative coefficient for $-5xy^2$ and $10xy$.

GCF = $-5xy$

Step 2. Rewrite each term of the polynomial as an equivalent product of $-5xy$ and a second factor.

$$-5xy^2 - 10xy$$
$$= \mathbf{-5xy} \cdot y - \mathbf{5xy} \cdot 2$$

Step 3. Use the distributive property to factor $-5xy$ from the resulting expression.

$$= \underbrace{\mathbf{-5xy} \cdot y + \mathbf{-5xy} \cdot 2}_{\text{Think}} = \mathbf{-5xy}(y+2)$$

To avoid sign errors, it is often helpful to mentally change a minus symbol to "+ −."

Step 4. Review the main steps.

$$-5xy^2 - 10xy = -5xy(y+2)$$

c. $-x - y$

Step 1. Insert the understood coefficients of 1.

$$-x - y$$
$$= -1x - 1y$$

Step 2. Use the distributive property to factor −1 from the polynomial.

$$= \underbrace{-1x + -1y}_{\text{Think}} = -1(x+y) = -(x+y)$$

Step 3. Review the main steps.

$$-x - y = -(x+y)$$

d. $-2x^3 + 4x - 8$

Step 1. Determine the GCF with a negative coefficient for $-2x^3$, $4x$, and -8.

GCF = -2

Step 2. Rewrite each term of the polynomial as an equivalent product of −2 and a second factor.

$$-2x^3 + 4x - 8$$
$$= \underbrace{\mathbf{-2} \cdot x^3 - \mathbf{2} \cdot -2x - \mathbf{2} \cdot 4}_{\text{Check signs}}$$

Step 3. Use the distributive property to factor -2 from the resulting expression.

$= \mathbf{-2}(x^3 - 2x + 4)$

Step 4. Review the main steps.

$$-2x^3 + 4x - 8 = -2\left(x^3 - 2x + 4\right)$$

A Quantity as a Common Factor

You might have a common quantity as a factor in the GCF.

Problem Factor.

a. $x(x - 1) + 2(x - 1)$

b. $a(c + d) + (c + d)$

c. $2x(x - 3) + 5(3 - x)$

Solution

a. $x(x - 1) + 2(x - 1)$

Step 1. Determine the GCF for $x(x - 1)$ and $2(x - 1)$.

GCF $= (x - 1)$

Step 2. Use the distributive property to factor $(x - 1)$ from the expression.

$$x(x-1)+2(x-1)$$
$$=(\mathbf{x}-\mathbf{1})(x+2)$$

b. $a(c + d) + (c + d)$

Step 1. Determine the GCF for $a(c + d)$ and $(c + d)$.

GCF $= (c + d)$

Step 2. Use the distributive property to factor $(c + d)$ from the expression.

$$a(c+d)+(c+d)$$
$$=\underbrace{a(c+d)+1(c+d)}_{\text{Think}}=(\mathbf{c}+\mathbf{d})(a+1)$$

$a(c + d) + (c + d) \neq (c + d)a$. Don't leave off the 1. Think of $(c + d)$ as $1(c + d)$.

c. $2x(x - 3) + 5(3 - x)$

Step 1. Because $(3 - x) = -1(x - 3)$, factor -1 from the second term.

$$2x(x-3)+5(3-x)$$
$$=2x(x-3)-5(x-3)$$

Step 2. Determine the GCF for $2x(x - 3)$ and $5(x - 3)$.

GCF = $(x - 3)$

Step 3. Use the distributive property to factor $(x - 3)$ from the expression.

$$2x(x-3)-5(x-3)$$
$$=(\boldsymbol{x}-\mathbf{3})(2x-5)$$

Factoring Four Terms

When you have four terms to factor, grouping the terms in pairs might yield a quantity as a common factor.

Problem Factor by grouping in pairs.

a. $x^2 + 2x + 3x + 6$

b. $ax + by + ay + bx$

Solution

a. $x^2 + 2x + 3x + 6$

Step 1. Group the terms in pairs that will yield a common factor.

$$x^2+2x+3x+6$$
$$=(x^2+2x)+(3x+6)$$

Step 2. Factor the common factor x out of the first term and the common factor 3 out of the second term.

$$=\boldsymbol{x}(x+2)+\mathbf{3}(x+2)$$

$(x^2 + 2x) + (3x + 6) \neq (x^2 + 2x)(3x + 6)$. These quantities are terms, not factors.

Step 3. Determine the GCF for $x(x + 2)$ and $3(x + 2)$.

GCF = $(x + 2)$

Step 4. Use the distributive property to factor $(x + 2)$ from the expression.

$$=x(x+2)+3(x+2)$$
$$=(\boldsymbol{x}+\mathbf{2})(x+3)$$

b. $ax + by + ay + bx$

Step 1. Rearrange the terms so that the first two terms have a common factor and the last two terms have a common factor, and then group the terms in pairs accordingly.

$$ax + by + ay + bx$$
$$= ax + bx + ay + by$$
$$= (ax + bx) + (ay + by)$$

Step 2. Factor the common factor x out of the first term and the common factor y out of the second term.

$$= \boldsymbol{x}(a + b) + \boldsymbol{y}(a + b)$$

Step 3. Determine the GCF for $x(a + b)$ and $y(a + b)$.

$$\text{GCF} = (a + b)$$

Step 4. Use the distributive property to factor $(a + b)$ from the expression.

$$x(a + b) + y(a + b)$$
$$= (\boldsymbol{a} + \boldsymbol{b})(x + y)$$

Factoring Quadratic Trinomials

When you have three terms to factor, you might have a *quadratic trinomial* of the form $ax^2 + bx + c$. It turns out that not all quadratic trinomials are factorable using real number coefficients, but many will factor. Those that do will factor as the product of two binomials.

> You should recall (from Chapter 9) that quadratic trinomials result when you multiply two binomials.

Two common methods for factoring $ax^2 + bx + c$ are *factoring by trial and error* and *factoring by grouping.*

Factoring by Trial and Error Using FOIL

When you factor by trial and error, it is very helpful to call to mind the FOIL method of multiplying two binomials. Here is an example.

$$(2x + 5)(3x + 4)$$
$$= \underbrace{2x \cdot 3x}_{\text{First}} + \underbrace{2x \cdot 4}_{\text{Outer}} + \underbrace{5 \cdot 3x}_{\text{Inner}} + \underbrace{5 \cdot 4}_{\text{Last}}$$
$$= 6x^2 + \underbrace{8x + 15x}_{\text{Middle terms}} + 20$$
$$= 6x^2 + 23x + 20$$

Your task when factoring $6x^2 + 23x + 20$ is to reverse the FOIL process to obtain $6x^2 + 23x + 20 = (2x + 5)(3x + 4)$ in factored form. As you work through the problems, you will find it useful to know the following: If the first and last terms of a factorable quadratic trinomial are positive, the signs of the second terms in the two binomial factors of the trinomial have the same sign as the middle term of the trinomial.

Problem Factor by trial and error.

a. $x^2 + 9x + 14$

b. $x^2 - 9x + 14$

c. $x^2 + 5x - 14$

d. $x^2 - 5x - 14$

e. $3x^2 + 5x - 2$

f. $4x^2 - 11x - 3$

Solution

a. $x^2 + 9x + 14$

Step 1. Because the expression has the form $ax^2 + bx + c$, look for two binomial factors.

$x^2 + 9x + 14 = (\quad)(\quad)$

Step 2. x^2 is the first term, so the first terms in the two binomial factors must be x.

$x^2 + 9x + 14 = (\boldsymbol{x}\quad)(\boldsymbol{x}\quad)$

Step 3. 14 is the last term, and it is positive, so the last terms in the two binomial factors have the same sign as 9, with a product of 14 and a sum of 9. Try 7 and 2 and check with FOIL.

$x^2 + 9x + 14 \stackrel{?}{=} (x + \mathbf{7})(x + \mathbf{2})$

Check: $(x+7)(x+2) = x^2 + 7x + 2x + 14 = x^2 + \underbrace{9x}_{\text{Correct}} + 14$

Step 4. Write the factored form.

$x^2 + 9x + 14 = (x + 7)(x + 2)$

b. $x^2 - 9x + 14$

Step 1. Because the expression has the form $ax^2 + bx + c$, look for two binomial factors.

$x^2 - 9x + 14 = (\quad)(\quad)$

Step 2. x^2 is the first term, so the first terms in the two binomial factors must be x.

$x^2 - 9x + 14 = (\mathbf{x}\quad)(\mathbf{x}\quad)$

Step 3. 14 is the last term, and it is positive, so the last terms in the two binomial factors have the same sign as −9, with a product of 14 and a sum of −9. Try −7 and −2 and check with FOIL.

$x^2 - 9x + 14 \stackrel{?}{=} (x - \mathbf{7})(x - \mathbf{2})$

Check: $(x-7)(x-2) = x^2 - 7x - 2x + 14 = x^2 \underbrace{-9x}_{\text{Correct}} + 14$

Step 4. Write the factored form.

$x^2 - 9x + 14 = (x-7)(x-2)$

c. $x^2 + 5x - 14$

Step 1. Because the expression has the form $ax^2 + bx + c$, look for two binomial factors.

$x^2 + 5x - 14 = (\quad)(\quad)$

Step 2. x^2 is the first term, so the first terms in the two binomial factors must be x.

$x^2 + 5x - 14 = (\mathbf{x}\quad)(\mathbf{x}\quad)$

Step 3. −14 is the last term, and it is negative, so the last terms in the two binomial factors have opposite signs with a product of −14 and a sum of 5. Try combinations of factors of −14 and check with FOIL.

Try $x^2 + 5x - 14 \stackrel{?}{=} (x + \mathbf{7})(x - \mathbf{2})$

Check: $(x+7)(x-2) = x^2 + 7x - 2x - 14 = x^2 + \underbrace{5x}_{\text{Correct}} - 14$

Step 4. Write the factored form.

$x^2 + 5x - 14 = (x+7)(x-2)$

d. $x^2 - 5x - 14$

Step 1. Because the expression has the form $ax^2 + bx + c$, look for two binomial factors.

$x^2 - 5x - 14 = (\quad)(\quad)$

Step 2. x^2 is the first term, so the first terms in the two binomial factors must be x.

$x^2 - 5x - 14 = (\mathbf{x}\quad)(\mathbf{x}\quad)$

Step 3. -14 is the last term, and it is negative, so the last terms in the two binomial factors have opposite signs with a product of -14 and a sum of -5. Try combinations of factors of -14 and check with FOIL.

Try $x^2 - 5x - 14 \stackrel{?}{=} (x - \mathbf{7})(x + \mathbf{2})$

Check: $(x-7)(x+2) = x^2 - 7x + 2x - 14 = x^2 \underbrace{-5x}_{\text{Correct}} - 14$

Step 4. Write the factored form.

$x^2 - 5x - 14 = (x-7)(x+2)$

e. $3x^2 + 5x - 2$

Step 1. Because the expression has the form $ax^2 + bx + c$, look for two binomial factors.

$3x^2 + 5x - 2 = (\quad)(\quad)$

Step 2. $3x^2$ is the first term, so the first terms in the two binomial factors must be $3x$ and x.

$3x^2 + 5x - 2 = (\mathbf{3x}\quad)(\mathbf{x}\quad)$

Step 3. -2 is the last term, and it is negative, so the last terms in the two binomial factors have opposite signs with a product of -2. Try combinations using the factors of -2 and check with FOIL until the middle term is correct.

Try $3x^2 + 5x - 2 \stackrel{?}{=} (3x - \mathbf{2})(x + \mathbf{1})$

Check: $(3x-2)(x+1) = 3x^2 + 3x - 2x - 2 = 3x^2 + \underbrace{x}_{\text{Wrong}} - 2$

Try $3x^2 + 5x - 2 \stackrel{?}{=} (3x + \mathbf{1})(x - \mathbf{2})$

Check: $(3x+1)(x-2) = 3x^2 - 6x + x - 2 = 3x^2 \underbrace{-5x}_{\text{Wrong}} - 2$

Try $3x^2 + 5x - 2 \stackrel{?}{=} (3x - \mathbf{1})(x + \mathbf{2})$

Check: $(3x-1)(x+2) = 3x^2 + 6x - x - 2 = 3x^2 + \underbrace{5x}_{\text{Correct}} - 2$

Step 4. Write the factored form.

$$3x^2 + 5x - 2 = (3x - 1)(x + 2)$$

f. $4x^2 - 11x - 3$

Step 1. Because the expression has the form $ax^2 + bx + c$, look for two binomial factors.

$$4x^2 - 11x - 3 = (\quad)(\quad)$$

Step 2. $4x^2$ is the first term, so the numerical coefficients of the first terms in the two binomial factors are factors of 4. The last term is -3, so the last terms of the two binomial factors have opposite signs with a product of -3. Try combinations of factors of 4 and -3 and check with FOIL until the middle term is correct.

Try $4x^2 - 11x - 3 \stackrel{?}{=} (\mathbf{2x} - \mathbf{3})(\mathbf{2x} + \mathbf{1})$

Check: $(2x - 3)(2x + 1) = 4x^2 + 2x - 6x - 3 = 4x^2 \underbrace{-4x}_{\text{Wrong}} - 3$

Try $4x^2 - 11x - 3 \stackrel{?}{=} (\mathbf{4x} - \mathbf{3})(x + \mathbf{1})$

Check: $(4x - 3)(x + 1) = 4x^2 + 4x - 3x - 3 = 4x^2 + \underbrace{x}_{\text{Wrong}} - 3$

Try $4x^2 - 11x - 3 \stackrel{?}{=} (\mathbf{4x} + \mathbf{1})(x - \mathbf{3})$

Check: $(4x + 1)(x - 3) = 4x^2 - 12x + x - 3 = 4x^2 \underbrace{-11x}_{\text{Correct}} - 3$

Step 3. Write the factored form.

$$4x^2 - 11x - 3 = (4x + 1)(x - 3)$$

As you can see, getting the middle term right is the key to a successful factorization of $ax^2 + bx + c$. You can shorten your checking time by simply using FOIL to compare the sum of the inner and outer products to the middle term of the trinomial.

Factoring by Grouping

When you factor $ax^2 + bx + c$ by grouping, you also guess and check, but in a different way than in the previous method.

Problem Factor by grouping.

a. $4x^2 - 11x - 3$

b. $9x^2 - 12x + 4$

Solution

a. $4x^2 - 11x - 3$

Step 1. Identify the coefficients a, b, and c and then find two factors of ac whose sum is b.

When you're identifying coefficients for $ax^2 + bx + c$, keep a – symbol with the number that follows it.

$a = 4, b = -11, \text{ and } c = -3$

$ac = 4 \cdot -3 = -12$

Two factors of -12 that sum to -11 are -12 and 1.

Step 2. Rewrite $4x^2 - 11x - 3$, replacing the middle term, $-11x$, with $-12x + 1x$.

$4x^2 - 11x - 3 = 4x^2 - \mathbf{12x + 1x} - 3$

Step 3. Group the terms in pairs that will yield a common factor.

$= (\mathbf{4x^2 - 12x}) + (\mathbf{1x - 3})$

Step 4. Factor the common factor $4x$ out of the first term and simplify the second term.

$= \mathbf{4x(x - 3) + (x - 3)}$

Step 5. Use the distributive property to factor $(x - 3)$ from the expression.

$= \mathbf{(x - 3)(4x + 1)}$

Step 6. Write the factored form.

$4x^2 - 11x - 3 = (x - 3)(4x + 1)$

b. $9x^2 - 12x + 4$

Step 1. Identify the coefficients a, b, and c and then find two factors of ac whose sum is b.

$a = 9, b = -12, \text{ and } c = 4$

$ac = 9 \cdot 4 = 36$

Two factors of 36 that sum to -12 are -6 and -6.

Step 2. Rewrite $9x^2 - 12x + 4$, replacing the middle term, $-12x$, with $-6x - 6x$.

$9x^2 - 12x + 4 = 9x^2 - \mathbf{6x - 6x} + 4$

Step 3. Group the terms in pairs that will yield a common factor.

$= \left(\mathbf{9x^2 - 6x}\right) - \left(\mathbf{6x \underset{\uparrow}{-} 4}\right)$

Check sign

Step 4. Factor the common factor $3x$ out of the first term and the common factor 2 out of the second term.

$= \mathbf{3x(3x-2)-2(3x-2)}$

Step 5. Use the distributive property to factor $(3x - 2)$ from the expression.

$= \mathbf{(3x - 2)(3x - 2)}$

Step 6. Write the factored form.

$9x^2 - 12x + 4 = (3x-2)^2$

Perfect Trinomial Squares

The trinomial $9x^2 - 12x + 4$, which equals $(3x - 2)^2$, is a perfect trinomial square. If you recognize that $ax^2 + bx + c$ is a perfect trinomial square, then you can factor it rather quickly. A trinomial is a perfect square if a and c are both positive and $|b| = 2\sqrt{a}\sqrt{c}$. The following problem illustrates the procedure.

Problem Factor.

a. $4x^2 - 20x + 25$

b. $x^2 + 6x + 9$

Solution

a. $4x^2 - 20x + 25$

Step 1. Identify the coefficients a, b, and c and check whether $|b| = 2\sqrt{a}\sqrt{c}$.

$a = 4$, $b = -20$, and $c = 25$

$|b| = |-20| = 20$ and $2\sqrt{a}\sqrt{c} = 2 \cdot \sqrt{4} \cdot \sqrt{25} = 2 \cdot 2 \cdot 5 = 20$

Thus, $4x^2 - 20x + 25$ is a perfect trinomial square.

Step 2. Indicate that $4x^2 - 20x + 25$ will factor as the square of a binomial.

$4x^2 - 20x + 25 = (\qquad)^2$

Step 3. Fill in the binomial. The first term is the square root of $4x^2$ and the last term is the square root of 25. The sign in the middle is the same as the sign of the middle term of the trinomial.

$4x^2 - 20x + 25 = \mathbf{(2x-5)^2}$ is the factored form.

b. $x^2 + 6x + 9$

Step 1. Identify the coefficients a, b, and c and check whether $|b| = 2\sqrt{a}\sqrt{c}$.

$a = 1, b = 6, \text{ and } c = 9$

$|b| = |6| = 6$ and $2\sqrt{a}\sqrt{c} = 2 \cdot \sqrt{1} \cdot \sqrt{9} = 2 \cdot 1 \cdot 3 = 6$

Thus, $x^2 + 6x + 9$ is a perfect trinomial square.

Step 2. Indicate that $x^2 + 6x + 9$ will factor as the square of a binomial.

$x^2 + 6x + 9 = (\quad)^2$

Step 3. Fill in the binomial. The first term is the square root of x^2 and the last term is the square root of 9. The sign in the middle is the same as the sign of the middle term of the trinomial.

$x^2 + 6x + 9 = \mathbf{(x + 3)^2}$ is the factored form.

Factoring Two Terms

When you have two terms to factor, consider these special binomial products from Chapter 9: the *difference of two squares*, the *difference of two cubes*, and the *sum of two cubes*.

$x^2 + y^2$, the *sum of two squares*, is not factorable (over the real numbers).

The *difference of two squares* has the form $x^2 - y^2$ (quantity squared minus quantity squared). You factor the difference of two squares like this:

$$x^2 - y^2 = (x + y)(x - y)$$

The *difference of two cubes* has the form $x^3 - y^3$ (quantity cubed minus quantity cubed). You factor the difference of two cubes like this:

$$x^3 - y^3 = (x - y)(x^2 + xy + y^2)$$

The *sum of two cubes* has the form $x^3 + y^3$ (quantity cubed plus quantity cubed). You factor the sum of two cubes like this:

$$x^3 + y^3 = (x + y)(x^2 - xy + y^2)$$

Problem Factor.

a. $9x^2 - 25y^2$

b. $x^2 - 1$

c. $x^2 + 4$

d. $8x^3 - 27$

e. $64a^3 + 125$

Solution

a. $9x^2 - 25y^2$

Step 1. Observe that the binomial has the form "quantity squared minus quantity squared," so it is the difference of two squares. Indicate that $9x^2 - 25y^2$ factors as the product of two binomials, one with a plus sign between the terms and the other with a minus sign between the terms.

$9x^2 - 25y^2 = (\quad + \quad)(\quad - \quad)$

Step 2. Fill in the terms of the binomials. The two first terms are the same, and they both equal $\sqrt{9x^2} = 3x$. The two last terms are the same, and they both equal $\sqrt{25y^2} = 5y$.

$9x^2 - 25y^2 = (\mathbf{3x + 5y})(\mathbf{3x - 5y})$ is the factored form.

b. $x^2 - 1$

Step 1. Observe that the binomial has the form "quantity squared minus quantity squared," so it is the difference of two squares. Indicate that $x^2 - 1$ factors as the product of two binomials, one with a plus sign between the terms and the other with a minus sign between the terms.

$x^2 - 1 = (\quad + \quad)(\quad - \quad)$

Step 2. Fill in the terms of the binomials. The two first terms are the same, and they both equal $\sqrt{x^2} = x$. The two last terms are the same, and they both equal $\sqrt{1} = 1$.

$x^2 - 1 = (\mathbf{x + 1})(\mathbf{x - 1})$ is the factored form.

c. $x^2 + 4$

Step 1. Observe that the binomial has the form "quantity squared *plus* quantity squared," so it is the *sum* of two squares, and thus is not factorable over the real numbers.

$x^2 + 4 \neq (x + 2)^2$. $(x + 2)^2 = x^2 + 4x + 4$, not $x^2 + 4$.

d. $8x^3 - 27$

Step 1. Observe that the binomial has the form "quantity cubed minus quantity cubed," so it is the difference of two cubes. Indicate that $8x^3 - 27$ factors as the product of a binomial and a trinomial. The binomial has a minus sign between the terms, and the trinomial has plus signs between the terms.

$8x^3 - 27 = (\quad - \quad)(\quad + \quad + \quad)$

Step 2. Fill in the terms of the binomial. The first term is $\sqrt[3]{8x^3} = 2x$, and the second term is $\sqrt[3]{27} = 3$.

$8x^3 - 27 = (\mathbf{2x - 3})(\quad + \quad + \quad)$

Step 3. Fill in the terms of the trinomial by using the terms of the binomial, $2x$ and 3, to obtain the terms. The first term is $(2x)^2 = 4x^2$, the second term is $2x \cdot 3 = 6x$, and the third term is $3^2 = 9$.

$8x^3 - 27 = (\mathbf{2x - 3})(\mathbf{4x^2 + 6x + 9})$ is the factored form.

> $4x^2 + 6x + 9 \neq (2x + 3)^2$. $(2x + 3)^2 = 4x^2 + \mathbf{12}x + 9$, not $4x^2 + 6x + 9$.

e. $64a^3 + 125$

Step 1. Observe that the binomial has the form "quantity cubed plus quantity cubed," so it is the sum of two cubes. Indicate that $64a^3 + 125$ factors as the product of a binomial and a trinomial. The binomial has a plus sign between the terms, and the trinomial has one minus sign on the middle term.

$64a^3 + 125 = (\quad + \quad)(\quad - \quad + \quad)$

Step 2. Fill in the terms of the binomial. The first term is $\sqrt[3]{64a^3} = 4a$, and the second term is $\sqrt[3]{125} = 5$.

$64a^3 + 125 = (\mathbf{4a + 5})(\quad - \quad + \quad)$

Step 3. Fill in the terms of the trinomial by using the terms of the binomial, $4a$ and 5, to obtain the terms. The first term is $(4a)^2 = 16a^2$, the second term is $4a \cdot 5 = 20a$, and the third term is $5^2 = 25$.

$64a^3 + 125 = (\mathbf{4a + 5})(\mathbf{16a^2 - 20a + 25})$

Guidelines for Factoring

Finally, here are some general guidelines for factoring of polynomials.

1. Count the number of terms.
2. If the expression has a GCF, factor out the GCF.
3. If there are two terms, check for a special binomial product.
4. If there are three terms, check for a quadratic trinomial.
5. If there are four terms, try grouping in pairs.
6. Check whether any previously obtained factor can be factored further.

Problem Factor completely.

a. $100x^4y^2z - 25x^2y^2z$

b. $x^2(x + y) + 2xy(x + y) + y^2(x + y)$

Solution

a. $100x^4y^2z - 25x^2y^2z$

Step 1. Factor out the GCF, $25x^2y^2z$.

$100x^4y^2z - 25x^2y^2z$

$= 25x^2y^2z \cdot 4x^2 - 25x^2y^2z \cdot 1$

$= \mathbf{25x^2y^2z}(4x^2 - 1)$

Step 2. Factor the difference of two squares, $4x^2 - 1$.

$\mathbf{25x^2y^2z(2x - 1)(2x + 1)}$ is the completely factored form.

b. $x^2(x + y) + 2xy\ (x + y) + y^2(x + y)$

Step 1. Factor out the GCF, $(x + y)$.

$x^2(x + y) + 2xy\ (x + y) + y^2(x + y)$

$= (x+y)(x^2+2xy+y^2)$

Step 2. Factor the perfect trinomial square, $x^2 + 2xy + y^2$, and simplify.

$(x+y)(x+y)^2 = (x+y)^3$ is the completely factored form.

Exercise 12

In 1–5, indicate whether the statement is true or false.

1. $\sqrt{64+25}=13$
2. $(x + 3)^2 = x^2 + 9$
3. $\dfrac{4+xy}{x}=4+y$
4. $\dfrac{1}{5+z}=\dfrac{1}{5}+\dfrac{1}{z}$
5. $\dfrac{1}{2^{-2}+3^{-2}}=2^2+3^2$

For 6–8, factor using a negative coefficient for the GCF.

6. $-a - b$
7. $-3x^2 + 6x - 9$
8. $3 - x$

For 9–24, factor completely.

9. $24x^9y^2 - 6x^6y^7z^4$
10. $-45x^2 + 5$
11. $a^3b - ab + b$
12. $14x + 7y$
13. $x(2x - 1) + 3(2x - 1)$
14. $y(a + b) + (a + b)$
15. $x(x - 3) + 2(3 - x)$
16. $cx + cy + ax + ay$
17. $x^2 - 3x - 4$
18. $x^2 - 49$
19. $6x^2 + x - 15$
20. $16x^2 - 25y^2$
21. $27x^3 - 64$
22. $8a^3 + 125b^3$
23. $2x^4y^2z^3 - 32x^2y^2z^3$
24. $a^2(a + b) - 2ab(a + b) + b^2(a + b)$

13

Rational Expressions

In this chapter, you apply your skills in factoring polynomials to the charge of simplifying rational expressions. A *rational expression* is an *algebraic fraction* that has a polynomial for its numerator and a polynomial for its denominator. For instance, $\frac{x^2-1}{x^2+2x+1}$ is a rational expression. Because division by 0 is undefined, you must exclude values for the variable or variables that would make the denominator polynomial sum to 0. For convenience, you can assume such values are excluded as you work through the problems in this chapter.

Reducing Algebraic Fractions to Lowest Terms

The following principle is fundamental to rational expressions.

Fundamental Principle of Rational Expressions

If P, Q, and R are polynomials, then $\frac{PR}{QR} = \frac{RP}{RQ} = \frac{P}{Q}$, provided neither Q nor R has a zero value.

The fundamental principle allows you to *reduce algebraic fractions to lowest terms* by dividing the numerator and denominator by the greatest common factor (GCF).

Before applying the fundamental principle of rational expressions, *always* make sure that the numerator and denominator contain only *factored* polynomials.

Problem Reduce to lowest terms.

a. $\dfrac{15x^5y^3z}{30x^5y^3}$

b. $\dfrac{6x}{2x}$

c. $\dfrac{x-3}{3-x}$

d. $\dfrac{3x}{3+x}$

e. $\dfrac{2x+6}{x^2+5x+6}$

f. $\dfrac{x^2-1}{x^2+2x+1}$

g. $\dfrac{x(a-b)+y(a-b)}{x+y}$

Solution

a. $\dfrac{15x^5y^3z}{30x^5y^3}$

Step 1. Determine the GCF for $15x^5y^3z$ and $30x^5y^3$.

$\text{GCF} = 15x^5y^3$

Step 2. Write the numerator and denominator as equivalent products with the GCF as one of the factors.

$$\frac{15x^5y^3z}{30x^5y^3} = \frac{15x^5y^3 \cdot z}{15x^5y^3 \cdot 2}$$

Step 3. Use the fundamental principle to reduce.

$$\frac{\cancel{15x^5y^3} \cdot z}{\cancel{15x^5y^3} \cdot 2} = \frac{z}{2}$$

b. $\frac{6x}{2x}$

Step 1. Determine the GCF for $6x$ and $2x$.

GCF $= 2x$

Step 2. Write the numerator and denominator as equivalent products with the GCF as one of the factors.

$$\frac{6x}{2x} = \frac{2x \cdot 3}{2x \cdot 1}$$

Step 3. Use the fundamental principle to reduce the fraction.

$$\frac{\cancel{2x} \cdot 3}{\cancel{2x} \cdot 1} = \frac{3}{1} = 3$$

c. $\frac{x-3}{3-x}$

Step 1. Factor -1 from the denominator polynomial, so that the x term will have a positive coefficient.

$$\frac{x-3}{3-x}$$
$$= \frac{x-3}{-1(-3+x)}$$
$$= \frac{x-3}{-1(x-3)}$$

Step 2. Determine the GCF for $x - 3$ and $-1(x - 3)$.

GCF $= (x - 3)$ (Enclose $x - 3$ in parentheses to emphasize it's a factor.)

Step 3. Write the numerator and denominator as equivalent products with the GCF as one of the factors.

$$\frac{x-3}{-1(-3+x)} = \frac{(x-3)}{-1(x-3)}$$

Step 4. Use the fundamental principle to reduce the fraction.

$$\frac{1\cancel{(x-3)}}{-1\cancel{(x-3)}} = -1$$

$\frac{(x-3)}{-1(x-3)} \neq \frac{0}{-1}$. Think of $(x - 3)$ as $1(x - 3)$.

d. $\frac{3x}{3+x}$

Step 1. Determine the GCF for $3x$ and $3 + x$.

GCF = 1, so $\frac{3x}{3+x}$ cannot be reduced further.

> $\frac{3x}{3+x} \neq \frac{x}{1+x}$. 3 is a factor of the numerator, but it is a *term* of the denominator. It is a mistake to divide out a term.

e. $\frac{2x+6}{x^2+5x+6}$

Step 1. Factor the numerator and denominator polynomials completely.

$$\frac{2x+6}{x^2+5x+6} = \frac{2(x+3)}{(x+2)(x+3)}$$

Step 2. Determine the GCF for $2(x + 3)$ and $(x + 2)(x + 3)$.

GCF = $(x + 3)$

Step 3. Use the fundamental principle to reduce the fraction.

$$\frac{2\cancel{(x+3)}}{(x+2)\cancel{(x+3)}} = \frac{2}{x+2}$$

> $\frac{2x+6}{x^2+5x+6} \neq \frac{2x}{x^2+5x}$. 6 is a common *term* in the numerator and denominator, not a factor. Only divide out factors.

f. $\frac{x^2-1}{x^2+2x+1}$

Step 1. Factor the numerator and denominator polynomials completely.

$$\frac{x^2-1}{x^2+2x+1} = \frac{(x+1)(x-1)}{(x+1)^2}$$

Step 2. Determine the GCF for $(x + 1)(x - 1)$ and $(x + 1)^2$.

GCF = $(x + 1)$

Step 3. Write the numerator and denominator as equivalent products with the GCF as one of the factors.

$$\frac{x^2-1}{x^2+2x+1} = \frac{(x+1)(x-1)}{(x+1)(x+1)}$$

Step 4. Use the fundamental principle to reduce the fraction.

$$\frac{\cancel{(x+1)}(x-1)}{\cancel{(x+1)}(x+1)} = \frac{x-1}{x+1}$$

g. $\dfrac{x(a-b)+y(a-b)}{x+y}$

Step 1. Factor the numerator and denominator polynomials completely.

$$\frac{x(a-b)+y(a-b)}{x+y} = \frac{(x+y)(a-b)}{(x+y)}$$

Step 2. Determine the GCF for $(x + y)(a - b)$ and $(x + y)$.

GCF = $(x + y)$

Step 3. Use the fundamental principle to reduce the fraction.

$$\frac{\cancel{(x+y)}(a-b)}{\cancel{(x+y)}} = \frac{a-b}{1} = a-b$$

Multiplying Algebraic Fractions

To multiply algebraic fractions, (1) factor all numerators and denominators completely, (2) divide numerators and denominators by their common factors (as in reducing), and (3) multiply the remaining numerator factors to get the numerator of the answer and multiply the remaining denominator factors to get the denominator of the answer.

Problem Find the product.

a. $\dfrac{x^2-2x+1}{x^2-4} \cdot \dfrac{3x-6}{x-1}$

b. $\dfrac{2x+4}{3-x} \cdot \dfrac{x^2-9}{x^2+5x+6}$

Solution

a. $\dfrac{x^2-2x+1}{x^2-4}\cdot\dfrac{3x-6}{x-1}$

Step 1. Factor all numerators and denominators completely.

$$\frac{x^2-2x+1}{x^2-4}\cdot\frac{3x-6}{x-1}$$

$$=\frac{(x-1)(x-1)}{(x+2)(x-2)}\cdot\frac{3(x-2)}{(x-1)}$$

When you are multiplying algebraic fractions, if a numerator or denominator does not factor, enclose it in parentheses. Forgetting the parentheses can lead to a mistake.

Step 2. Divide out common numerator and denominator factors.

$$=\frac{\cancel{(x-1)}(x-1)}{(x+2)\cancel{(x-2)}}\cdot\frac{3\cancel{(x-2)}}{\cancel{(x-1)}}$$

Be careful! Only divide out factors.

Step 3. Multiply the remaining numerator factors to get the numerator of the answer and multiply the remaining denominator factors to get the denominator of the answer.

$$=\frac{3(x-1)}{(x+2)}$$

Step 4. Review the main results.

$$\frac{x^2-2x+1}{x^2-4}\cdot\frac{3x-6}{x-1}$$

$$=\frac{\cancel{(x-1)}(x-1)}{(x+2)\cancel{(x-2)}}\cdot\frac{3\cancel{(x-2)}}{\cancel{(x-1)}}=\frac{3(x-1)}{(x+2)}$$

When you multiply algebraic fractions, you can leave your answer in factored form. Always double-check to make sure it is in completely reduced form.

b. $\dfrac{2x+4}{3-x}\cdot\dfrac{x^2-9}{x^2+5x+6}$

Step 1. Factor all numerators and denominators completely.

$$\frac{2x+4}{3-x}\cdot\frac{x^2-9}{x^2+5x+6}$$

$$=\frac{2(x+2)}{-1(x-3)}\cdot\frac{(x+3)(x-3)}{(x+2)(x+3)}$$

Write all polynomial factors with the variable terms first, so that you can easily recognize common factors.

Step 2. Divide out common numerator and denominator factors.

$$= \frac{2\cancel{(x+2)}}{-1\cancel{(x-3)}} \cdot \frac{\cancel{(x+3)}\cancel{(x-3)}}{\cancel{(x+2)}\cancel{(x+3)}}$$

Step 3. Multiply the remaining numerator factors to get the numerator of the answer, and multiply the remaining denominator factors to get the denominator of the answer.

$$= \frac{2}{-1} = -2$$

Step 4. Review the main results.

$$\frac{2x+4}{3-x} \cdot \frac{x^2-9}{x^2+5x+6} = \frac{2\cancel{(x+2)}}{-1\cancel{(x-3)}} \cdot \frac{\cancel{(x+3)}\cancel{(x-3)}}{\cancel{(x+2)}\cancel{(x+3)}} = -2$$

Dividing Algebraic Fractions

To divide algebraic fractions, multiply the first algebraic fraction by the reciprocal of the second algebraic fraction (the divisor).

Problem Find the quotient: $\frac{x^2-2x+1}{x^2-x-6} \div \frac{x^2-3x+2}{x^2-4}$.

Solution

Step 1. Change the problem to multiplication by the reciprocal of the divisor.

$$\frac{x^2-2x+1}{x^2-x-6} \div \frac{x^2-3x+2}{x^2-4}$$

$$= \frac{x^2-2x+1}{x^2-x-6} \cdot \frac{x^2-4}{x^2-3x+2}$$

Step 2. Factor all numerators and denominators completely.

$$= \frac{(x-1)(x-1)}{(x-3)(x+2)} \cdot \frac{(x+2)(x-2)}{(x-1)(x-2)}$$

Step 3. Divide out common numerator and denominator factors.

$$= \frac{\cancel{(x-1)}(x-1)}{(x-3)\cancel{(x+2)}} \cdot \frac{\cancel{(x+2)}\cancel{(x-2)}}{\cancel{(x-1)}\cancel{(x-2)}}$$

Step 4. Multiply the remaining numerator factors to get the numerator of the answer, and multiply the remaining denominator factors to get the denominator of the answer.

$$= \frac{(x-1)}{(x-3)}$$

Step 5. Review the main results.

$$\frac{x^2-2x+1}{x^2-x-6} \div \frac{x^2-3x+2}{x^2-4} = \frac{x^2-2x+1}{x^2-x-6} \cdot \frac{x^2-4}{x^2-3x+2}$$

$$= \frac{\cancel{(x-1)}(x-1)}{(x-3)\cancel{(x+2)}} \cdot \frac{\cancel{(x+2)}\cancel{(x-2)}}{\cancel{(x-1)}\cancel{(x-2)}} = \frac{(x-1)}{(x-3)}$$

Adding (or Subtracting) Algebraic Fractions, Like Denominators

To add (or subtract) algebraic fractions that have like denominators, place the sum (or difference) of the numerators over the common denominator. Simplify and reduce to lowest terms, if needed.

Problem Compute as indicated.

a. $\frac{x+2}{x-3} + \frac{2x-11}{x-3}$

b. $\frac{5x^2}{4(x+1)} - \frac{4x^2+1}{4(x+1)}$

Solution

a. $\frac{x+2}{x-3} + \frac{2x-11}{x-3}$

Step 1. Indicate the sum of the numerators over the common denominator.

$$\frac{x+2}{x-3} + \frac{2x-11}{x-3}$$

$$= \frac{(x+2)+(2x-11)}{x-3}$$

Step 2. Find the sum of the numerators.

$$= \frac{x+2+2x-11}{x-3}$$

$$= \frac{3x-9}{x-3}$$

Step 3. Reduce to lowest terms.

$$= \frac{3(x-3)}{(x-3)} = \frac{3\cancel{(x-3)}}{\cancel{(x-3)}} = \frac{3}{1} = 3$$

Step 4. Review the main results.

$$\frac{x+2}{x-3} + \frac{2x-11}{x-3} = \frac{3x-9}{x-3} = \frac{3\cancel{(x-3)}}{\cancel{(x-3)}} = \frac{3}{1} = 3$$

b. $\frac{5x^2}{4(x+1)} - \frac{4x^2+1}{4(x+1)}$

Step 1. Indicate the difference of the numerators over the common denominator.

$$\frac{5x^2}{4(x+1)} - \frac{4x^2+1}{4(x+1)}$$

$$= \frac{(5x^2)-(4x^2+1)}{4(x+1)}$$

When subtracting algebraic fractions, it is important that you enclose the numerator of the second fraction in parentheses because you want to subtract the *entire numerator*, not just the first term.

Step 2. Find the difference of the numerators.

$$= \frac{5x^2-4x^2-1}{4(x+1)}$$

$$= \frac{x^2-1}{4(x+1)}$$

Step 3. Reduce to lowest terms.

$$= \frac{x^2-1}{4(x+1)} = \frac{\cancel{(x+1)}(x-1)}{4\cancel{(x+1)}} = \frac{(x-1)}{4}$$

Step 4. Review the main results.

$$\frac{5x^2}{4(x+1)}-\frac{4x^2+1}{4(x+1)}=\frac{5x^2-4x^2-1}{4(x+1)}=\frac{x^2-1}{4(x+1)}=\frac{\cancel{(x+1)}(x-1)}{4\cancel{(x+1)}}=\frac{(x-1)}{4}$$

Adding (or Subtracting) Algebraic Fractions, Unlike Denominators

To add (or subtract) algebraic fractions that have unlike denominators, (1) factor each denominator completely; (2) find the least common denominator (LCD), which is the product of each prime factor the *highest* number of times it is a factor in any one denominator; (3) using the fundamental principle, write each algebraic fraction as an equivalent fraction having the common denominator as a denominator; and (4) add (or subtract) as for like denominators.

Note: A *prime factor* is one that cannot be factored further.

Problem Compute as indicated.

a. $\frac{3x}{x^2-4}+\frac{x}{x-2}$

b. $\frac{2x-1}{x-3}-\frac{x}{2x+2}$

Solution

a. $\frac{3x}{x^2-4}+\frac{x}{x-2}$

Step 1. Factor each denominator completely.

$$\frac{3x}{x^2-4}+\frac{x}{x-2}$$

$$=\frac{3x}{(x+2)(x-2)}+\frac{x}{(x-2)}$$

Step 2. Find the LCD.

$$\text{LCD}=(x+2)(x-2)$$

Step 3. Write each algebraic fraction as an equivalent fraction having the common denominator as a denominator.

$$= \frac{3x}{(x+2)(x-2)} + \frac{x\cdot(x+2)}{(x-2)\cdot(x+2)}$$

$$= \frac{3x}{(x+2)(x-2)} + \frac{x^2+2x}{(x-2)(x+2)}$$

Step 4. Add as for like denominators.

$$= \frac{(3x)+\left(x^2+2x\right)}{(x+2)(x-2)}$$

$$= \frac{3x+x^2+2x}{(x+2)(x-2)}$$

$$= \frac{x^2+5x}{(x+2)(x-2)}$$

$$= \frac{x(x+5)}{(x+2)(x-2)}$$

Step 5. Review the main results.

$$\frac{3x}{x^2-4} + \frac{x}{x-2} = \frac{3x}{(x+2)(x-2)} + \frac{x\cdot(x+2)}{(x-2)\cdot(x+2)} = \frac{3x+x^2+2x}{(x+2)(x-2)}$$

$$= \frac{x^2+5x}{(x+2)(x-2)} = \frac{x(x+5)}{(x+2)(x-2)}$$

b. $\frac{2x-1}{x-3} - \frac{x}{2x+2}$

Step 1. Factor each denominator completely.

$$\frac{2x-1}{x-3} - \frac{x}{2x+2}$$

$$= \frac{2x-1}{(x-3)} - \frac{x}{2(x+1)}$$

Step 2. Find the LCD.

$$\text{LCD} = 2(x-3)(x+1)$$

Step 3. Write each algebraic fraction as an equivalent fraction having the common denominator as a denominator.

$$=\frac{(2x-1)\cdot 2(x+1)}{(x-3)\cdot 2(x+1)}-\frac{x\cdot(x-3)}{2(x+1)\cdot(x-3)}$$

$$=\frac{4x^2+2x-2}{2(x-3)(x+1)}-\frac{x^2-3x}{2(x-3)(x+1)}$$

Step 4. Subtract as for like denominators.

$$=\frac{(4x^2+2x-2)}{2(x-3)(x+1)}-\frac{(x^2-3x)}{2(x-3)(x+1)}$$

$$=\frac{(4x^2+2x-2)-(x^2-3x)}{2(x-3)(x+1)}$$

$$=\frac{4x^2+2x-2-x^2+3x}{2(x-3)(x+1)}$$

$$=\frac{3x^2+5x-2}{2(x-3)(x+1)}$$

$$=\frac{(3x-1)(x+2)}{2(x-3)(x+1)}$$

Step 5. Review the main results.

$$\frac{2x-1}{x-3}-\frac{x}{2x+2}=\frac{2x-1}{(x-3)}-\frac{x}{2(x+1)}$$

$$=\frac{(2x-1)\cdot 2(x+1)}{(x-3)\cdot 2(x+1)}-\frac{x\cdot(x-3)}{2(x+1)\cdot(x-3)}$$

$$=\frac{4x^2+2x-2}{2(x-3)(x+1)}-\frac{x^2-3x}{2(x-3)(x+1)}=\frac{4x^2+2x-2-x^2+3x}{2(x-3)(x+1)}$$

$$=\frac{3x^2+5x-2}{2(x-3)(x+1)}$$

$$=\frac{(3x-1)(x+2)}{2(x-3)(x+1)}$$

Complex Fractions

A *complex fraction* is a fraction that has fractions in its numerator, denominator, or both. One way you can simplify a complex fraction is to interpret the fraction bar of the complex fraction as meaning division.

Problem Simplify $\dfrac{\frac{1}{x}+\frac{1}{y}}{\frac{1}{x}-\frac{1}{y}}$.

Solution

Step 1. Write the complex fraction as a division problem.

$$\dfrac{\frac{1}{x}+\frac{1}{y}}{\frac{1}{x}-\frac{1}{y}}$$

$$=\left(\frac{1}{x}+\frac{1}{y}\right)\div\left(\frac{1}{x}-\frac{1}{y}\right)$$

Step 2. Perform the indicated addition and subtraction.

$$=\left(\frac{y+x}{xy}\right)\div\left(\frac{y-x}{xy}\right)$$

Step 3. Multiply by the reciprocal of the divisor.

$$=\frac{(y+x)}{xy}\cdot\frac{xy}{(y-x)}$$

$$=\frac{(y+x)}{\cancel{xy}}\cdot\frac{\cancel{xy}}{(y-x)}$$

$$=\frac{(y+x)}{(y-x)}$$

Step 4. Review the main results.

$$\frac{\frac{1}{x}+\frac{1}{y}}{\frac{1}{x}-\frac{1}{y}}=\left(\frac{1}{x}+\frac{1}{y}\right)\div\left(\frac{1}{x}-\frac{1}{y}\right)=\left(\frac{y+x}{xy}\right)\div\left(\frac{y-x}{xy}\right)=$$

$$\frac{(y+x)}{\cancel{xy}}\cdot\frac{\cancel{xy}}{(y-x)}=\frac{(y+x)}{(y-x)}$$

Another way you can simplify a complex fraction is to multiply its numerator and denominator by the LCD of all the fractions in its numerator and denominator.

Problem Simplify: $\dfrac{\frac{1}{x}+\frac{1}{y}}{\frac{1}{x}-\frac{1}{y}}$.

Solution

Step 1. Multiply the numerator and denominator by the LCD of all the fractions.

$$\frac{\frac{1}{x}+\frac{1}{y}}{\frac{1}{x}-\frac{1}{y}}$$

$$=\frac{xy\left(\frac{1}{x}+\frac{1}{y}\right)}{xy\left(\frac{1}{x}-\frac{1}{y}\right)}$$

$$=\frac{xy\cdot\frac{1}{x}+xy\cdot\frac{1}{y}}{xy\cdot\frac{1}{x}-xy\cdot\frac{1}{y}}$$

$$=\frac{y+x}{y-x}$$

Exercise 13

For 1–10, reduce to lowest terms.

1. $\dfrac{18x^3y^4z^2}{54x^3z^2}$

2. $\dfrac{15y}{3y}$

3. $\dfrac{x-5}{5-x}$

4. $\dfrac{4a}{4+a}$

5. $\dfrac{2x-6}{x^2-5x+6}$

6. $\dfrac{x^2-4}{x^2+4x+4}$

7. $\dfrac{x(a+b)+y(a+b)}{x+y}$

8. $\dfrac{7x}{35x-14}$

9. $\dfrac{4x^2y-4xy-24y}{2x^2-18}$

10. $\dfrac{x-y}{x^3-y^3}$

For 11–15, compute as indicated.

11. $\dfrac{x^2-4x+4}{x^2-9}\cdot\dfrac{2x-6}{x-2}$

12. $\dfrac{x-1}{2x-1}\div\dfrac{x+1}{4x-2}$

13. $\dfrac{2}{x^2-2x-3}+\dfrac{4}{x-3}$

14. $\dfrac{2x}{x^2-14x+49}-\dfrac{1}{x-7}$

15. $\dfrac{\frac{2}{3x}}{\frac{1}{x+1}}$

14

Solving Linear Equations and Inequalities

A linear equation in one variable, say x, has the standard form $ax + b = c$, $a \neq 0$, where a, b, and c are constants. For example, $3x - 7 = 14$ is a linear equation in standard form. An equation has two sides. The expression on the left side of the equal sign is the *left side* of the equation, and the expression on the right side of the equal sign is the *right side* of the equation.

Solving One-Variable Linear Equations

To *solve a linear equation* that has one variable x means to find a numerical value for x that makes the equation true. An equation is true when the left side has the same value as the right side. When you solve an equation, you undo what has been done to x until you get an expression like this: x = a number. As you proceed, you exploit the fact that addition and subtraction undo each other; and, similarly, multiplication and division undo one another.

The goal in solving a linear equation is to get the variable by itself on only one side of the equation and with a coefficient of 1 (usually understood).

You solve an equation using the properties of real numbers and simple algebraic tools. An equation is like a balance scale. To keep the equation in balance, when you do something to one side of the equation, you must do to the same thing to the other side of the equation.

Tools for Solving Linear Equations

Add the same number to both sides.
Subtract the same number from both sides.
Multiply both sides by the same *nonzero* number.
Divide both sides by the same *nonzero* number.

It is important to remember that when you are solving an equation, you must *never* multiply or divide both sides by 0.

Problem Solve the equation.

a. $5x + 9 = 3x - 1$

b. $4(x - 6) = 40$

c. $-3x - 7 = 14$

d. $3x - 2 = 7 - 2x$

e. $\frac{x-3}{2} = \frac{2x+4}{5}$

Solution

a. $5x + 9 = 3x - 1$

Step 1. The variable appears on both sides of the equation, so subtract $3x$ from the right side to remove it from that side. To maintain balance, subtract $3x$ from the left side, too.

$5x + 9 - \mathbf{3x} = 3x - 1 - \mathbf{3x}$

Step 2. Simplify both sides by combining like variable terms.

$\mathbf{2x} + 9 = -1$

Step 3. 9 is added to the variable term, so subtract 9 from both sides.

$2x + 9 - \mathbf{9} = -1 - \mathbf{9}$

Step 4. Simplify both sides by combining constant terms.

$\mathbf{2x = -10}$

Step 5. You want the coefficient of x to be 1, so divide both sides by 2.

$\frac{2x}{\mathbf{2}} = \frac{-10}{\mathbf{2}}$

Step 6. Simplify.

$\mathbf{x = -5}$

Step 7. Check your answer by substituting -5 for x in the original equation, $5x + 9 = 3x - 1$

Substitute -5 for x on the left side of the equation: $5x + 9 = 5(-5) + 9 = -25 + 9 = -16$. Similarly, on the right side, you have $3x - 1 = 3(-5) - 1 = -15 - 1 = -16$. Both sides equal -16, so -5 is the correct answer.

b. $4(x - 6) = 40$

Step 1. Use the distributive property to remove parentheses.

$\mathbf{4x - 24} = 40$

Step 2. 24 is subtracted from the variable term, so add 24 to both sides.

$4x - 24 + \mathbf{24} = 40 + \mathbf{24}$

Step 3. Simplify both sides by combining constant terms.

$\mathbf{4x = 64}$

Step 4. You want the coefficient of x to be 1, so divide both sides by 4.

$$\frac{4x}{\mathbf{4}} = \frac{64}{\mathbf{4}}$$

Step 5. Simplify.

$\mathbf{x = 16}$

Step 6. Check your answer by substituting 16 for x in the original equation, $4(x - 6) = 40$.

Substitute 16 for x on the left side of the equation: $4(x - 6) = 4(16 - 6) = 4(10) = 40$. On the right side, you have 40 as well. Both sides equal 40, so 16 is the correct answer.

c. $-3x - 7 = 14$

Step 1. 7 is subtracted from the variable term, so add 7 to both sides.

$-3x - 7 + \mathbf{7} = 14 + \mathbf{7}$

Step 2. Simplify both sides by combining constant terms.

$\mathbf{-3x = 21}$

Step 3. You want the coefficient of x to be 1, so divide both sides by -3.

$$\frac{-3x}{\mathbf{-3}} = \frac{21}{\mathbf{-3}}$$

Step 4. Simplify.

$\mathbf{x = -7}$

Step 5. Check your answer by substituting –7 for x in the original equation, $-3x - 7 = 14$.

Substitute –7 for x on the left side of the equation: $-3x - 7 = -3(-7) - 7 = 21 - 7 = 14$. On the right side, you have 14 as well. Both sides equal 14, so –7 is the correct answer.

d. $3x - 2 = 7 - 2x$

Step 1. The variable appears on both sides of the equation, so add $2x$ to the right side to remove it from that side. To maintain balance, add $2x$ to the left side, too.

$3x - 2 + \mathbf{2x} = 7 - 2x + \mathbf{2x}$

Step 2. Simplify both sides by combining like variable terms.

$\mathbf{5x} - 2 = 7$

Step 3. 2 is subtracted from the variable term, so add 2 to both sides.

$5x - 2 + \mathbf{2} = 7 + \mathbf{2}$

Step 4. Simplify both sides by combining constant terms.

$\mathbf{5x = 9}$

Step 5. You want the coefficient of x to be 1, so divide both sides by 5.

$$\frac{5x}{\mathbf{5}} = \frac{9}{\mathbf{5}}$$

Step 6. Simplify.

$\mathbf{x = 1.8}$

Step 7. Check your answer by substituting 1.8 for x in the original equation, $3x - 2 = 7 - 2x$.

Substitute 1.8 for x on the left side of the equation: $3x - 2 = 3(1.8) - 2 = 5.4 - 2 = 3.4$. Similarly, on the right side, you have $7 - 2x = 7 - 2(1.8) = 7 - 3.6 = 3.4$. Both sides equal 3.4, so 1.8 is the correct answer.

e. $\frac{x-3}{2}=\frac{2x+4}{5}$

Step 1. Eliminate fractions by multiplying both sides by 10, the least common multiple of 2 and 5. Write 10 as $\frac{10}{1}$ to avoid errors.

$$\frac{\mathbf{10}}{1}\cdot\frac{x-3}{2}=\frac{\mathbf{10}}{1}\cdot\frac{2x+4}{5}$$

Step 2. Simplify.

$$\frac{{}^{5}\cancel{10}}{1}\cdot\frac{x-3}{\cancel{2}_{1}}=\frac{{}^{2}\cancel{10}}{1}\cdot\frac{2x+4}{\cancel{5}_{1}}$$

$$\mathbf{5(x-3)=2(2x+4)}$$

$$\mathbf{5x-15=4x+8}$$

Step 3. The variable appears on both sides of the equation, so subtract $4x$ from the right side to remove it from that side. To maintain balance, subtract $4x$ from the left side, too.

$$5x-15-\mathbf{4x}=4x+8-\mathbf{4x}$$

Step 4. Simplify both sides by combining variable terms.

$$\mathbf{x}-15=8$$

Step 5. 15 is subtracted from the variable term, so add 15 to both sides.

$$x-15+\mathbf{15}=8+\mathbf{15}$$

Step 6. Simplify both sides by combining constant terms.

$$\mathbf{x=23}$$

Step 7. Check your answer by substituting 23 for x in the original equation, $\frac{x-3}{2}=\frac{2x+4}{5}$.

Substitute 23 for x on the left side of the equation: $\frac{x-3}{2}=\frac{23-3}{2}$ $=\frac{20}{2}=10$. Similarly, on the right side, you have $\frac{2x+4}{5}=\frac{2(23)+4}{5}$ $=\frac{46+4}{5}=\frac{50}{5}=10$. Both sides equal 10, so 23 is the correct answer.

Solving Two-Variable Linear Equations for a Specific Variable

You can use the procedures for solving a linear equation in one variable x to solve a two-variable linear equation, such as $6x + 2y = 10$, for one of the variables in terms of the other variable. As you solve for the variable of interest, you simply treat the other variable as you would a constant. Often, you need to solve for y to facilitate the graphing of an equation. (See Chapter 17 for a fuller discussion of this topic.) Here is an example.

Problem Solve $6x + 2y = 10$ for y.

Solution

Step 1. $6x$ is added to the variable term $2y$, so subtract $6x$ from both sides.

$6x + 2y - \mathbf{6x} = 10 - \mathbf{6x}$

When you are solving $6x + 2y = 10$ for y, treat $6x$ as if it were a constant.

Step 2. Simplify.

$\mathbf{2y = 10 - 6x}$

Step 3. You want the coefficient of y to be 1, so divide both sides by 2.

$$\frac{2y}{\mathbf{2}} = \frac{10 - 6x}{\mathbf{2}}$$

$\frac{10-6x}{2} \neq 5-6x$. You must divide *both* terms of the numerator by 2.

Step 4. Simplify.

$\mathbf{y = 5 - 3x}$

Solving Linear Inequalities

If you replace the equal sign in a linear equation with $<$, $>$, $\leq$, or $\geq$, the result is a linear inequality. You solve linear inequalities just about the same way you solve equations. There is just one important difference. When you multiply or divide both sides of an inequality by a *negative* number, you must *reverse* the direction of the inequality symbol. To help you understand why you must do this, consider the two numbers, 8 and 2. You know that $8 > 2$ is a true inequality because 8 is to the right of 2 on the number line, as shown in Figure 14.1.

If you multiply both sides of the inequality $8 > 2$ by a negative number, say, -1, you must reverse the direction of the inequality so that you will still

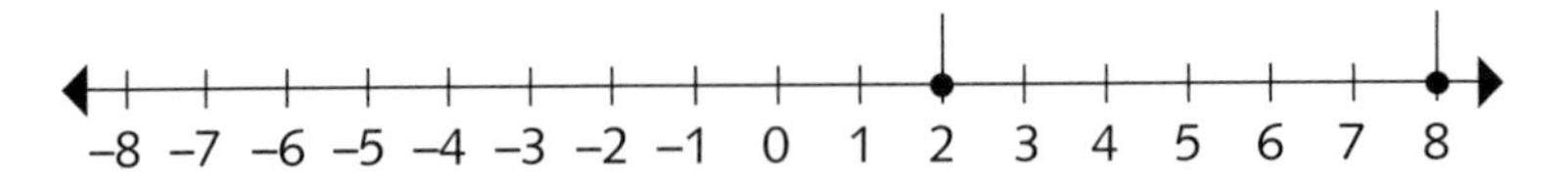

Figure 14.1 The numbers 2 and 8 on the number line

have a true inequality, namely, $-8 < -2$. You can verify that $-8 < -2$ is a true inequality by observing that -2 is to the right of -8 on the number line as shown in Figure 14.2.

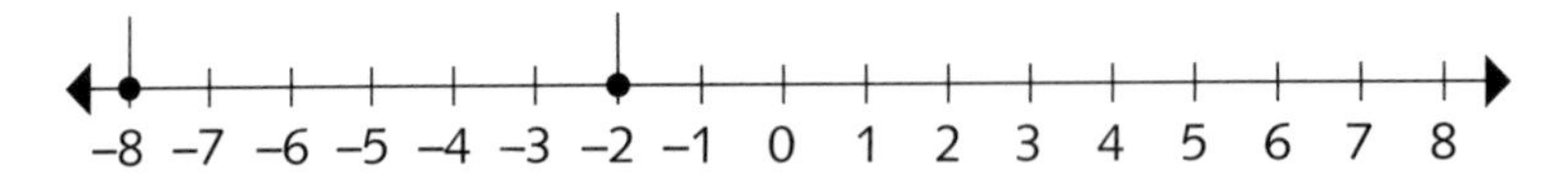

Figure 14.2 The numbers −8 and −2 on the number line

If you neglect to reverse the direction of the inequality symbol after multiplying both sides of $8 > 2$ by -1, you get the false inequality $-8 > -2$.

Problem Solve the inequality.

a. $5x + 6 < 3x - 2$

b. $4(x - 6) \geq 44$

c. $-3x - 7 > 14$

Solution

a. $5x + 6 < 3x - 2$

Step 1. The variable appears on both sides of the inequality, so subtract $3x$ from the right side to remove it from that side. To maintain balance, subtract $3x$ from the left side, too.

$5x + 6 - \mathbf{3x} < 3x - 2 - \mathbf{3x}$

Step 2. Simplify both sides by combining like variable terms.

$\mathbf{2x} + 6 < -2$

Step 3. 6 is added to the variable term, so subtract 6 from both sides.

$2x + 6 - \mathbf{6} < -2 - \mathbf{6}$

When solving an inequality, do *not* reverse the direction of the inequality symbol because of subtracting the same number from both sides.

Step 4. Simplify both sides by combining constant terms.

$\mathbf{2x < -8}$

Step 5. You want the coefficient of x to be 1, so divide both sides by 2.

$$\frac{2x}{\mathbf{2}} < \frac{-8}{\mathbf{2}}$$

When solving an inequality, do *not* reverse the direction of the inequality symbol because of dividing both sides by a positive number.

Step 6. Simplify.

$\mathbf{x < -4}$ is the answer.

b. $4(x - 6) \geq 44$

Step 1. Use the distributive property to remove parentheses.

$\mathbf{4x - 24} \geq 44$

Step 2. 24 is subtracted from the variable term, so add 24 to both sides.

$4x - 24 \mathbf{+ 24} \geq 44 \mathbf{+ 24}$

When solving an inequality, do *not* reverse the direction of the inequality because of adding the same number to both sides.

Step 3. Simplify both sides by combining constant terms.

$\mathbf{4x \geq 68}$

Step 4. You want the coefficient of x to be 1, so divide both sides by 4.

$$\frac{4x}{\mathbf{4}} \geq \frac{68}{\mathbf{4}}$$

Step 5. Simplify.

$\mathbf{x \geq 17}$ is the answer.

c. $-3x - 7 > 14$

Step 1. 7 is subtracted from the variable term, so add 7 to both sides.

$-3x - 7 \mathbf{+ 7} > 14 \mathbf{+ 7}$

Step 2. Simplify both sides by combining constant terms.

$\mathbf{-3x > 21}$

Step 3. You want the coefficient of x to be 1, so divide both sides by -3 and *reverse the direction of the inequality because you divided by a negative number.*

$$\frac{-3x}{\mathbf{-3}} < \frac{21}{\mathbf{-3}}$$

When solving an inequality, remember to reverse the direction of the inequality when you divide both sides by the same negative number.

Step 4. Simplify.

$\boldsymbol{x} < \mathbf{-7}$ is the answer.

Exercise 14

For 1–5, solve the equation for x.

1. $x - 7 = 11$
2. $6x - 3 = 13$
3. $x + 3(x - 2) = 2x - 4$
4. $\dfrac{x+3}{5} = \dfrac{x-1}{2}$
5. $3x + 2 = 6x - 4$
6. Solve for y: $-12x + 6y = 9$

For 7–10, solve the inequality for x.

7. $-x + 9 < 0$
8. $3x + 2 > 6x - 4$
9. $3x - 2 \leq 7 - 2x$
10. $\dfrac{x+3}{5} \geq \dfrac{x-1}{2}$

15

Solving Quadratic Equations

Quadratic equations in the variable x can always be put in the standard form $ax^2 + bx + c = 0$, $a \neq 0$. This type of equation is *always* solvable for the variable x, and each result is a *root* of the quadratic equation. In one instance the solution will yield only complex number roots. This case will be singled out in the discussion that follows. You will get a feel for the several ways of solving quadratic equations by starting with simple equations and working up to the most general equations. The discussion will be restricted to real number solutions. When instructions are given to solve the system, then you are to find all *real* numbers x that will make the equation true. These values (if any) are the real roots of the quadratic equation.

Solving Quadratic Equations of the Form $ax^2 + c = 0$

Normally, the first step in solving a quadratic equation is to put it in standard form. However, if there is no x term, that is, if the coefficient b is 0, then you have a simple way to solve such quadratic equations.

Problem Solve $x^2 = -4$.

Solution

Step 1. Because the square of a real number is never negative, there is no real number solution to the system.

Problem Solve $x^2 = 7$.

Solution

Step 1. Solve for x^2.

Step 2. Because both sides are nonnegative, take the square root of both sides.

$$\sqrt{x^2} = \sqrt{7}$$

Recall that the principal square root is always nonnegative and the equation $\sqrt{x^2} = |x|$ was discussed at length in Chapter 3.

Step 3. Simplify and write the solution.

$$|x| = \sqrt{7}$$

Thus, $x = \sqrt{7}$ or $x = -\sqrt{7}$.

A solution such as $x = \sqrt{7}$ or $x = -\sqrt{7}$ is usually written $x = \pm\sqrt{7}$.

As you gain more experience, the solution of an equation such as $\sqrt{x^2} = k$, $k \geq 0$, can be considerably shortened if you remember that $\sqrt{x^2} = |x|$ and apply that idea mentally. You can write the solution immediately as $x = \pm\sqrt{k}$.

Problem Solve $x^2 - 6 = 0$.

Solution

Step 1. Solve for x^2 to obtain the form for a quick solution.

$$x^2 = 6$$

Step 2. Write the solution.

The solution is $x = \pm\sqrt{6}$.

Problem Solve $3x^2 = 48$.

Solution

Step 1. Solve for x^2 to obtain the form for a quick solution.

$$x^2 = 16$$

Step 2. Write the solution.

The solution is $x = \pm 4$.

When the coefficient b of a quadratic equation is not 0, the quick solution method does not work. Instead, you have three common methods for solving the equation: (1) by factoring, (2) by completing the square, and (3) by using the quadratic formula.

Solving Quadratic Equations by Factoring

When you solve quadratic equations by factoring, you use the following property of 0.

Zero Factor Property

If the product of two numbers is 0, then at least one of the numbers is 0.

Problem Solve by factoring.

a. $x^2 + 2x = 0$

b. $x^2 + x = 6$

c. $x^2 - 4x = -4$

Solution

a. $x^2 + 2x = 0$

Step 1. Put the equation in standard form.

$x^2 + 2x = 0$ is in standard form because only a zero term is on the right side.

Step 2. Use the distributive property to factor the left side of the equation.

$x(x + 2) = 0$

Step 3. Use the zero factor property to separate the factors.

Thus, $x = 0$ or $x + 2 = 0$.

Step 4. Solve the resulting linear equations.

The solution is $x = 0$ or $x = -2$.

b. $x^2 + x = 6$

Step 1. Put the equation in standard form.

$$x^2 + x - 6 = 0$$

Step 2. Factor.

$$(x-2)(x+3) = 0$$

Step 3. Use the zero factor property to separate the factors.

Thus, $x - 2 = 0$ or $x + 3 = 0$.

Step 4. Solve the resulting linear equations.

The solution is $x = 2$ or $x = -3$.

c. $x^2 - 4x = -4$

Step 1. Put the equation in standard form.

$$x^2 - 4x + 4 = 0$$

Step 2. Factor.

$$(x-2)(x-2) = 0$$

$$(x-2)^2 = 0$$

Step 3. Write the quick solution.

$$(x-2)^2 = 0$$

$$x - 2 = \pm 0 = 0$$

The solution is $x = 2$.

Solving Quadratic Equations by Completing the Square

You also can use the technique of completing the square to solve quadratic equations. This technique starts off differently in that you do not begin by putting the equation in standard form.

Problem Solve $x^2 - 2x = 6$ by completing the square.

Solution

Step 1. Complete the square on the left side by adding the square of $\frac{1}{2}$ the coefficient of x, being sure to maintain the balance of the equation by adding the same quantity to the right side.

$$x^2 - 2x + \quad = 6 +$$

$$x^2 - 2x + 1 = 6 + 1$$

Step 2. Factor the left side.

$$(x+1)(x+1) = 7$$

$$(x+1)^2 = 7$$

Step 3. Solve using the quick solution method.

$$x + 1 = \pm\sqrt{7}$$

$$x = -1 \pm \sqrt{7}$$

Thus, $x = -1 + \sqrt{7}$ or $x = -1 - \sqrt{7}$.

Solving Quadratic Equations by Using the Quadratic Formula

Having illustrated several useful approaches, it turns out there is one technique that will *always* solve *any* quadratic equation that is in standard form. This method is solving by using the quadratic formula.

Quadratic Formula

The solution of the quadratic equation $ax^2 + bx + c = 0$ is given by the formula $x = \frac{-b \pm \sqrt{b^2 - 4ac}}{2a}$. The term under the radical, $b^2 - 4ac$, is called the discriminant of the quadratic equation.

If $b^2 - 4ac = 0$, there is only one root for the equation. If $b^2 - 4ac > 0$, there are two real number roots. And if $b^2 - 4ac < 0$, there is no real number solution. In the latter case, both roots are complex numbers because this solution involves the square root of a negative number.

Problem Solve by using the quadratic formula.

a. $3x^2 - 2x + 11 = 0$

b. $2x^2 + 2x - 5 = 0$

c. $x^2 - 6x + 9 = 0$

Solution

a. $3x^2 - 2x + 11 = 0$

Step 1. Identify the coefficients a, b, and c and then use the quadratic formula.

When you're identifying coefficients for $ax^2 + bx + c = 0$, *keep a – symbol with the number that follows it.*

$a = 3$, $b = -2$, and $c = 11$

$$x = \frac{-b \pm \sqrt{b^2 - 4ac}}{2a} = \frac{-(-2) \pm \sqrt{(-2)^2 - 4(3)(11)}}{2(3)}$$

$$= \frac{2 \pm \sqrt{4 - 132}}{6}$$

$$= \frac{2 \pm \sqrt{-128}}{6}$$

Step 2. State the solution.

Because the discriminant is negative there is no real number solution for $3x^2 - 2x + 11 = 0$.

b. $2x^2 + 2x - 5 = 0$

Step 1. Identify the coefficients a, b, and c and then use the quadratic formula.

$a = 2$, $b = 2$, and $c = -5$

$$x = \frac{-b \pm \sqrt{b^2 - 4ac}}{2a} = \frac{-(2) \pm \sqrt{(2)^2 - 4(2)(-5)}}{2(2)}$$

$$= \frac{-2 \pm \sqrt{4 + 40}}{4}$$

$$= \frac{-2 \pm \sqrt{44}}{4}$$

$$= \frac{-2 \pm \sqrt{4(11)}}{4}$$

$$= \frac{-2 \pm 2\sqrt{11}}{4} = \frac{2(-1 \pm \sqrt{11})}{4}$$

$$= \frac{-1 \pm \sqrt{11}}{2}$$

Step 2. State the solution.

The solution is $x = \dfrac{-1+\sqrt{11}}{2}$ or $x = \dfrac{-1-\sqrt{11}}{2}$.

c. $x^2 - 6x + 9 = 0$

Step 1. Identify the coefficients a, b, and c and then use the quadratic formula.

$a = 1$, $b = -6$, and $c = 9$

$$x = \frac{-b \pm \sqrt{b^2 - 4ac}}{2a} = \frac{-(-6) \pm \sqrt{(-6)^2 - 4(1)(9)}}{2(1)}$$

$$= \frac{6 \pm \sqrt{36 - 36}}{2}$$

$$= \frac{6 \pm \sqrt{0}}{2} = \frac{6}{2} = 3$$

Step 2. State the solution.

The solution is $x = 3$.

Exercise 15

1. Solve $x^2 - x - 6 = 0$ by factoring.
2. Solve $x^2 + 6x = -5$ by completing the square.
3. Solve $3x^2 - 5x + 1 = 0$ by using the quadratic formula.

For 4–10, solve by any method.

4. $x^2 - 6 = 8$
5. $x^2 - 3x + 2 = 0$
6. $9x^2 + 18x - 17 = 0$
7. $6x^2 - 12x + 7 = 0$
8. $x^2 - 10x = -25$
9. $-x^2 = -9$
10. $6x^2 = x + 2$

16

The Cartesian Coordinate Plane

In this chapter, you learn about the Cartesian coordinate plane.

Definitions for the Plane

The *Cartesian coordinate plane* is defined by two real number lines, one horizontal and one vertical, intersecting at right angles at their zero points (see Figure 16.1). The two real number lines are the *coordinate axes*. The *horizontal axis*, commonly called the *x-axis*, has positive direction to the right, and the *vertical axis*, commonly referred to as the *y-axis*, has positive direction upward. The two axes determine a plane. Their point of intersection is called the *origin*.

Ordered Pairs in the Plane

In the (Cartesian) coordinate plane, you identify each point P in the plane by an *ordered pair* (x, y) of real numbers x and y, called its *coordinates*. The ordered pair (0, 0) names the origin. An ordered pair of numbers is written in a definite order so that one number is first and the other second. The first number is the *x-coordinate*, and the second number is the *y-coordinate* (see Figure 16.2). The order in the ordered pair (x,y) that corresponds to a point P is important. The absolute value of the first coordinate, x, is the perpendicular horizontal distance (right or left) of the point P from the y-axis. If x is positive, P is to the right of the y-axis; if x is negative, it is to the left of the y-axis. The absolute value of the second coordinate, y, is the perpendicular

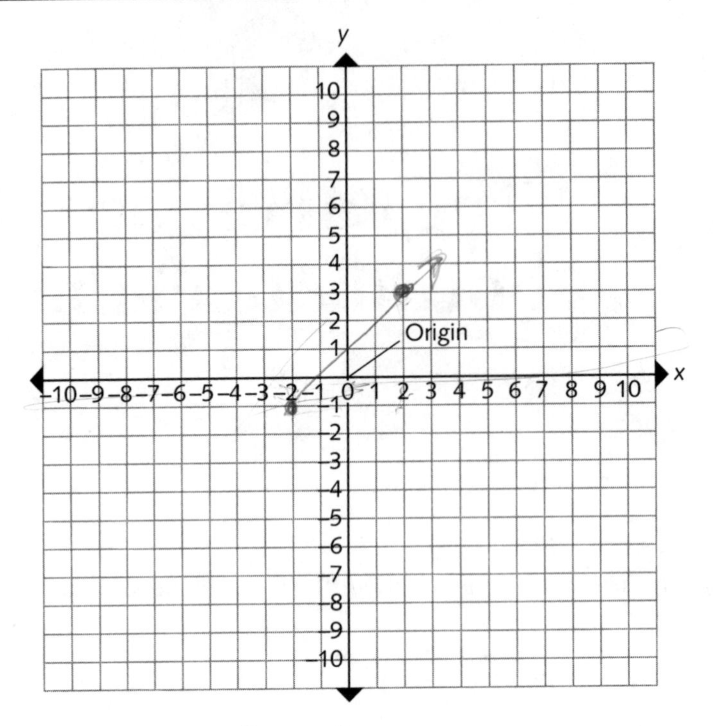

Figure 16.1 Cartesian coordinate plane

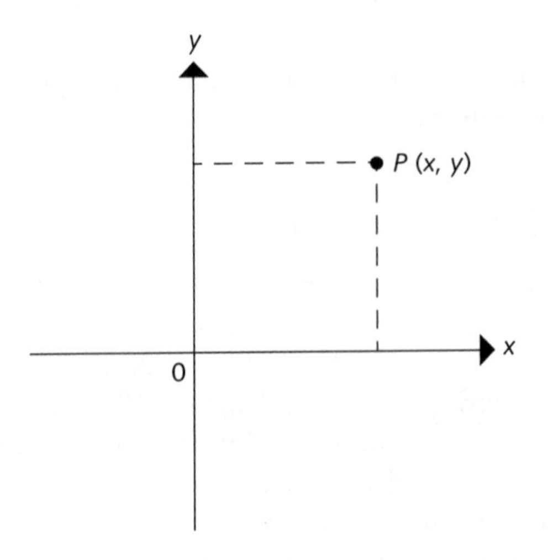

Figure 16.2 Point P in a Cartesian coordinate plane

vertical distance (up or down) of the point P from the x-axis. If y is positive, P is above the x-axis; if y is negative, it is below the x-axis.

Problem Name the ordered pair of integers corresponding to point A in the coordinate plane shown.

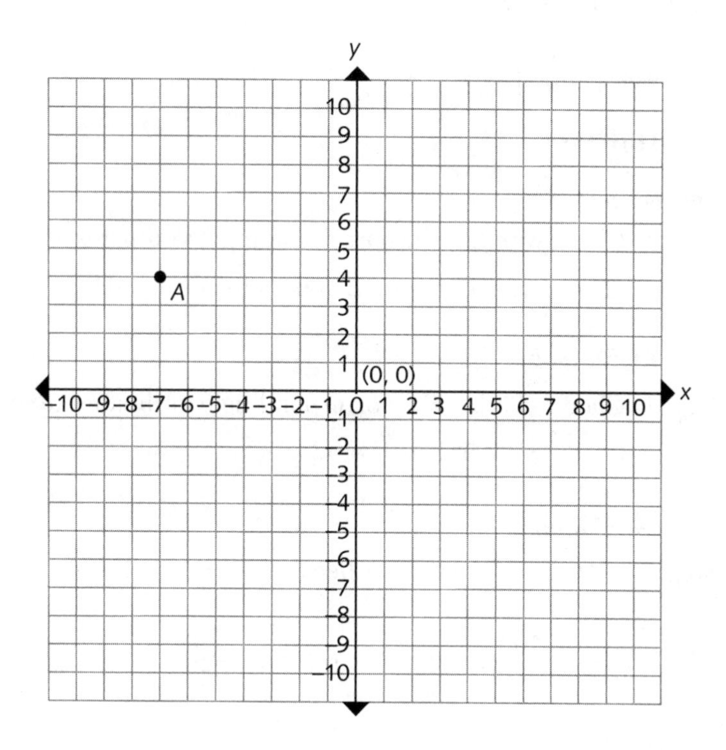

Solution

Step 1. Determine the x-coordinate of A.

The point A is 7 units to the left of the y-axis, so it has x-coordinate -7.

Step 2. Determine the y-coordinate of A.

The point A is 4 units above the x-axis, so it has y-coordinate 4.

Step 3. Name the ordered pair corresponding to point A.

$(-7, 4)$ is the ordered pair corresponding to point A.

Two ordered pairs are equal if and only if their corresponding coordinates are equal; that is, $(a, b) = (c, d)$ if and only if $a = c$ and $b = d$.

Problem State whether the two ordered pairs are equal. Explain your answer.

a. (2, 7), (7, 2)

b. (–3, 5), (3, 5)

c. (–4, –1), (4, 1)

d. $(6, 2), \left(6, \frac{10}{5}\right)$

Solution

a. (2, 7), (7, 2)

Step 1. Check whether the corresponding coordinates are equal.

$(2, 7) \neq (7, 2)$ because 2 and 7 are not equal.

b. (–3, 5), (3, 5)

Step 1. Check whether the corresponding coordinates are equal.

$(-3, 5) \neq (3, 5)$ because $-3 \neq 3$.

c. (–4, –1), (4, 1)

Step 1. $(-4, -1) \neq (4, 1)$ either because $-4 \neq 4$ or because $-1 \neq 1$.

d. $(6, 2) = \left(6, \frac{10}{5}\right)$ because $6 = 6$ and $2 = \frac{10}{5}$.

Quadrants of the Plane

The axes divide the Cartesian coordinate plane into four *quadrants*. The quadrants are numbered with Roman numerals—I, II, III, and IV—beginning in the upper right and going around counterclockwise, as shown in Figure 16.3.

Don't forget that the quadrants are numbered *counterclockwise*.

In quadrant I, both coordinates are positive; in quadrant II, the x-coordinate is negative, and the y-coordinate is positive; in quadrant III, both coordinates are negative; and in quadrant IV, the x-coordinate is positive, and the y-coordinate is negative. Points that have 0 as one or both of the coordinates are on the axes. If the x-coordinate is 0, the point lies on the y-axis. If the y-coordinate is 0, the point lies on the x-axis. If both coordinates of a point are 0, the point is at the origin.

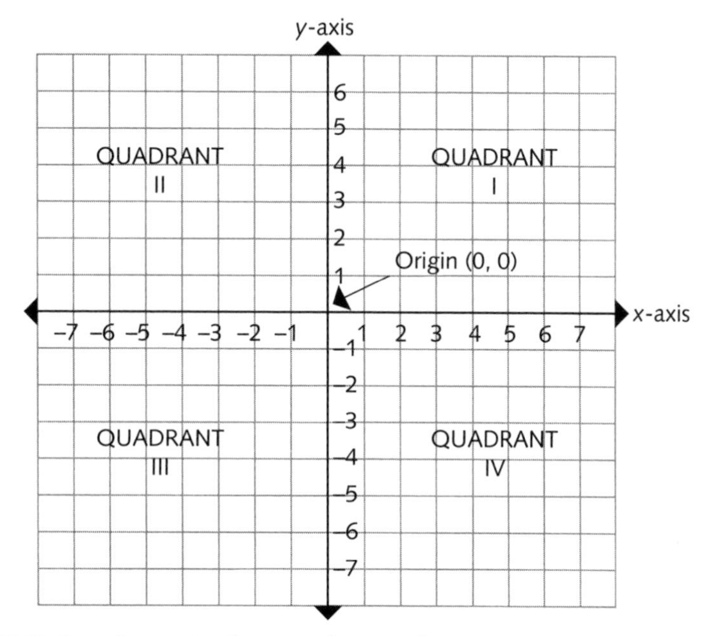

Figure 16.3 Quadrants in the coordinate plane

Problem Identify the quadrant in which the point lies.

a. (4, −8)

b. (1, 6)

c. (−8, −3)

d. (−4, 2)

Solution

a. (4, −8)

Step 1. Note the signs of x and y.

x is positive and y is negative.

Step 2. Identify the quadrant in which the point lies.

Because x is positive and y is negative, (4, −8) lies in quadrant IV.

b. (1, 6)

Step 1. Note the signs of x and y.

x is positive and y is positive.

Step 2. Identify the quadrant in which the point lies.

Because x is positive and y is positive, (1, 6) lies in quadrant I.

c. (−8, −3)

Step 1. Note the signs of x and y.

x is negative and y is negative.

Step 2. Identify the quadrant in which the point lies.

Because x is negative and y is negative, (−8, −3) lies in quadrant III.

d. (−4, 2)

Step 1. Note the signs of x and y.

x is negative and y is positive.

Step 2. Identify the quadrant in which the point lies.

Because x is negative and y is positive, (−4, 2) lies in quadrant II.

Finding the Distance Between Two Points in the Plane

If you have two points in a coordinate plane, you can find the distance between them using the formula given here.

Distance Between Two Points

The distance d between two points (x_1, y_1) and (x_2, y_2) in a coordinate plane is given by

$$\text{Distance} = d = \sqrt{(x_2 - x_1)^2 + (y_2 - y_1)^2}$$

To avoid careless errors when using the distance formula, enclose substituted *negative* values in parentheses.

Problem Find the distance between the points (−1, 4) and (5, −3).

Solution

Step 1. Specify (x_1, y_1) and (x_2, y_2) and identify values for x_1, y_1, x_2, and y_2.

Let $(x_1, y_1) = (-1, 4)$ and $(x_2, y_2) = (5, -3)$. Then $x_1 = -1$, $y_1 = 4$, $x_2 = 5$, and $y_2 = -3$.

Step 2. Evaluate the formula for the values from step 1.

$$d = \sqrt{(x_2 - x_1)^2 + (y_2 - y_1)^2} = \sqrt{(5-(-1))^2 + ((-3)-4)^2}$$

$$= \sqrt{(5+1)^2 + (-3-4)^2} = \sqrt{(6)^2 + (-7)^2} = \sqrt{36+49} = \sqrt{85}$$

Step 3. State the distance.

The distance between (–1, 4) and (5, –3) is $\sqrt{85}$ units.

Finding the Midpoint Between Two Points in the Plane

You can find the midpoint between two points using the following formula.

Midpoint Between Two Points

The midpoint between two points (x_1, y_1) and (x_2, y_2) in a coordinate plane is the point with coordinates

$$\left(\frac{x_1 + x_2}{2}, \frac{y_1 + y_2}{2}\right)$$

When you use the midpoint formula, be sure to put plus signs, not minus signs, between the two *x* values and the two *y* values.

Problem Find the midpoint between (–1, 4) and (5, –3).

Solution

Step 1. Specify (x_1, y_1) and (x_2, y_2) and identify values for x_1, y_1, x_2, and y_2.

Let $(x_1, y_1) = (-1, 4)$ and $(x_2, y_2) = (5, -3)$. Then $x_1 = -1$, $y_1 = 4$, $x_2 = 5$, and $y_2 = -3$.

Step 2. Evaluate the formula for the values from step 1.

$$\text{Midpoint} = \left(\frac{x_1 + x_2}{2}, \frac{y_1 + y_2}{2}\right) = \left(\frac{-1+5}{2}, \frac{4-3}{2}\right) = \left(\frac{4}{2}, \frac{1}{2}\right) = \left(2, \frac{1}{2}\right)$$

Step 3. State the midpoint.

The midpoint between (–1, 4) and (5, –3) is $\left(2, \frac{1}{2}\right)$.

Finding the Slope of a Line Through Two Points in the Plane

When you have two distinct points in a coordinate plane, you can construct the line through the two points. The *slope* describes the steepness or slant (if any) of the line. To calculate the slope of a line, use the following formula.

Slope of a Line Through Two Points

The slope m of a line through two distinct points, (x_1, y_1) and (x_2, y_2), is given by

$$\text{Slope} = m = \frac{y_2 - y_1}{x_2 - x_1}, \text{ provided } x_1 \neq x_2$$

When you use the slope formula, be sure to subtract the coordinates in the same order in both the numerator and the denominator. That is, if y_2 is the first term in the numerator, then x_2 must be the first term in the denominator. It is also a good idea to enclose substituted *negative* values in parentheses to guard against careless errors.

From the formula, you can see that the slope is the ratio of the change in vertical coordinates (the *rise*) to the change in horizontal coordinates (the *run*). Thus, slope $= \frac{\text{rise}}{\text{run}}$. Figure 16.4 illustrates the rise and run for the slope of the line through points $P_1(x_1, y_1)$ and $P_2(x_2, y_2)$.

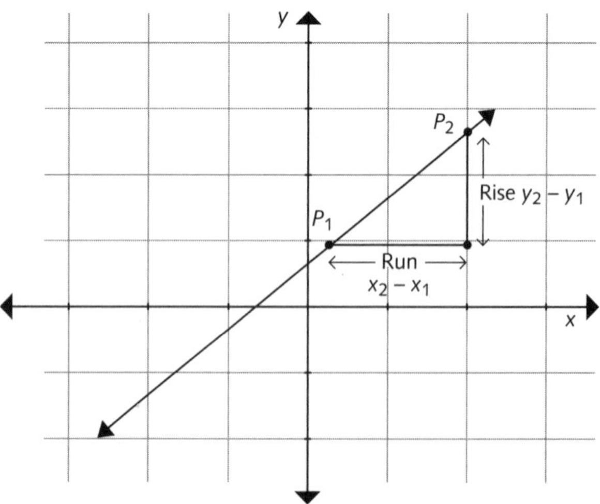

Figure 16.4 Rise and run

You will find it helpful to know that lines that slant upward from left to right have positive slopes, and lines that slant downward from left to right have negative slopes. Also, horizontal lines have zero slope, but the slope for vertical lines is undefined.

Problem Find the slope of the line through the points shown.

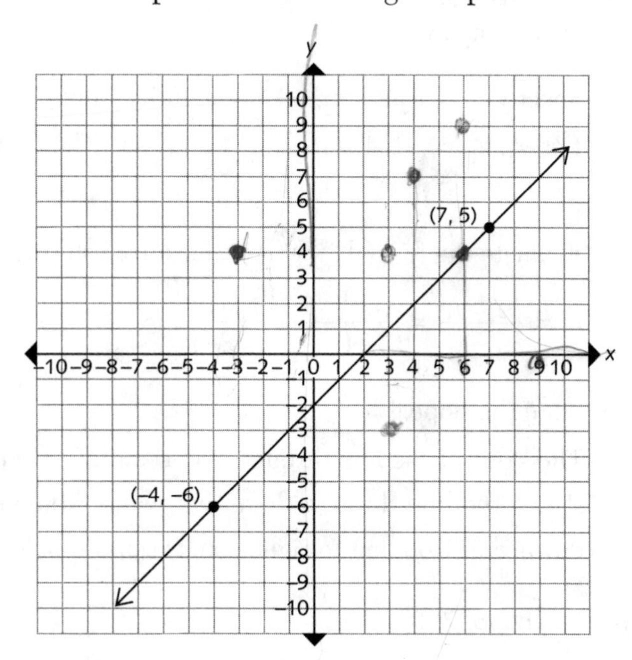

Solution

Step 1. Specify (x_1, y_1) and (x_2, y_2) and identify values for x_1, y_1, x_2, and y_2.

Let $(x_1, y_1) = (7, 5)$ and $(x_2, y_2) = (-4, -6)$. Then $x_1 = 7$, $y_1 = 5$, $x_2 = -4$, and $y_2 = -6$.

Step 2. Evaluate the formula for the values from step 1.

$$m = \frac{y_2 - y_1}{x_2 - x_1} = \frac{(-6) - 5}{(-4) - 7} = \frac{-6 - 5}{-4 - 7} = \frac{-11}{-11} = 1$$

Step 3. State the slope.

The slope of the line that passes through the points (7, 5) and (−4, −6) is 1. *Note:* The line slants upward from left to right—so the slope should be positive.

Problem Find the slope of the line through the two points.

a. (−1, 4) and (5, −3)

b. (−6, 7) and (5, 7)

c. (5, 8) and (5, −3)

Solution

a. (−1, 4) and (5, −3)

Step 1. Specify (x_1, y_1) and (x_2, y_2) and identify values for x_1, y_1, x_2, and y_2.

Let $(x_1, y_1) = (-1, 4)$ and $(x_2, y_2) = (5, -3)$. Then $x_1 = -1$, $y_1 = 4$, $x_2 = 5$, and $y_2 = -3$.

Step 2. Evaluate the formula for the values from step 1.

$$m = \frac{y_2 - y_1}{x_2 - x_1} = \frac{(-3)-4}{5-(-1)} = \frac{-3-4}{5+1} = \frac{-7}{6} = -\frac{7}{6}$$

Step 3. State the slope.

The slope of the line through (−1, 4) and (5, −3) is $-\frac{7}{6}$. *Note:* If you sketch the line through these two points, you will see that it slants downward from left to right—so its slope should be negative.

b. (−6, 7) and (5, 7)

Step 1. Specify (x_1, y_1) and (x_2, y_2) and identify values for x_1, y_1, x_2, and y_2.

Let $(x_1, y_1) = (-6, 7)$ and $(x_2, y_2) = (5, 7)$. Then $x_1 = -6$, $y_1 = 7$, $x_2 = 5$, and $y_2 = 7$.

Step 2. Evaluate the formula for the values from step 1.

$$m = \frac{y_2 - y_1}{x_2 - x_1} = \frac{7-7}{5-(-6)} = \frac{7-7}{5+6} = \frac{0}{11} = 0$$

Step 3. State the slope.

The slope of the line that contains (−6, 7) and (5, 7) is 0. *Note:* If you sketch the line through these two points, you will see that it is a horizontal line—so the slope should be 0.

c. (5, 8) and (5, −3)

Step 1. Specify (x_1, y_1) and (x_2, y_2) and identify values for x_1, y_1, x_2, and y_2.

Let $(x_1, y_1) = (5, 8)$ and $(x_2, y_2) = (5, -3)$. Then $x_1 = 5$, $y_1 = 8$, $x_2 = 5$, and $y_2 = -3$.

Step 2. Evaluate the formula for the values from step 1.

$$m = \frac{y_2 - y_1}{x_2 - x_1} = \frac{(-3)-8}{5-5} = \frac{-3-8}{5-5} = \frac{-11}{0} = \text{undefined}$$

Step 3. State the slope.

The slope of the line that contains (5, 8) and (5, –3) is undefined. *Note:* If you sketch the line through these two points, you will see that it is a vertical line—so the slope should be undefined.

Slopes of Parallel and Perpendicular Lines

It is useful to know the following:

If two lines are parallel, their slopes are equal; if two lines are perpendicular, their slopes are negative reciprocals of each other.

Problem Find the indicated slope.

a. Find the slope m_1 of a line that is parallel to the line through (–3, 4) and (–1, –2).

b. Find the slope m_2 of a line that is perpendicular to the line through (–3, 4) and (–1, –2).

Solution

a. Find the slope m_1 of a line that is parallel to the line through (–3, 4) and (–1, –2).

Step 1. Determine a strategy.

Because two parallel lines have equal slopes, m_1 will equal the slope m of the line through (–3, 4) and (–1, –2); that is, $m_1 = m$.

Step 2. Find m.

$$m = \frac{y_2 - y_1}{x_2 - x_1} = \frac{(-2)-4}{(-1)-(-3)} = \frac{-2-4}{-1+3} = \frac{-6}{2} = -3$$

Step 3. Determine m_1.

$m_1 = m = -3$

b. Find the slope m_2 of a line that is perpendicular to the line through (–3, 4) and (–1, –2).

Step 1. Determine a strategy.

Because the slopes of two perpendicular lines are negative reciprocals of each other, m_2 will equal the negative reciprocal of the slope m of the line through (–3, 4) and (–1, –2); that is, $m_2 = -\frac{1}{m}$.

Step 2. Find m.

$$m = \frac{y_2 - y_1}{x_2 - x_1} = \frac{(-2)-4}{(-1)-(-3)} = \frac{-2-4}{-1+3} = \frac{-6}{2} = -3$$

Step 3. Determine m_2.

$$m_2 = -\frac{1}{m} = -\frac{1}{-3} = \frac{1}{3}$$

Exercise 16

For 1–6, indicate whether the statement is true or false.

1. The intersection of the coordinate axes is the origin.
2. $(2, 3) = (3, 2)$
3. $\left(\frac{2}{3}, \frac{1}{2}\right) = \left(\frac{4}{6}, \frac{5}{10}\right)$
4. The point $\left(\frac{3}{4}, -5\right)$ is in quadrant II.
5. The point $\left(-\frac{\sqrt{2}}{2}, -\frac{\sqrt{2}}{2}\right)$ is in quadrant III.
6. The point (5, 0) is in quadrant I.

For 7–14, fill in the blank to make a true statement.

7. The change in *y*-coordinates between two points on a line is the ________.
8. The change in *x*-coordinates between two points on a line is the ________.
9. Lines that slant downward from left to right have ________ slopes.
10. Lines that slant upward from left to right have ________ slopes.
11. Horizontal lines have ________ slope.
12. The slope of a line that is parallel to a line that has slope $\frac{2}{3}$ is ________.
13. The slope of a line that is perpendicular to a line that has slope $\frac{3}{4}$ is ________.
14. The slope of a vertical line is ________.
15. Name the ordered pair of integers corresponding to point *K* in the following coordinate plane.

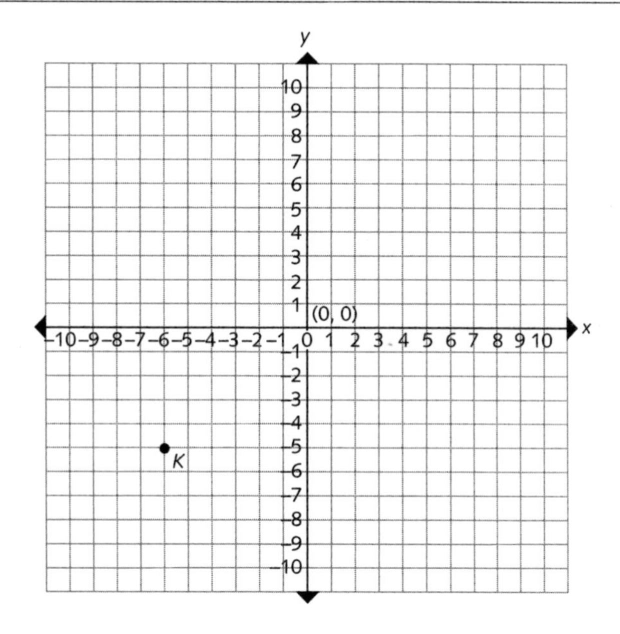

16. Find the distance between the points (1, 4) and (5, 7).
17. Find the distance between the points (–2, 5) and (4, –1).
18. Find the midpoint between the points (–2, 5) and (4, –1).
19. Find the slope of the line through (–2, 5) and (4, –1).
20. Find the slope of the line through the points shown.

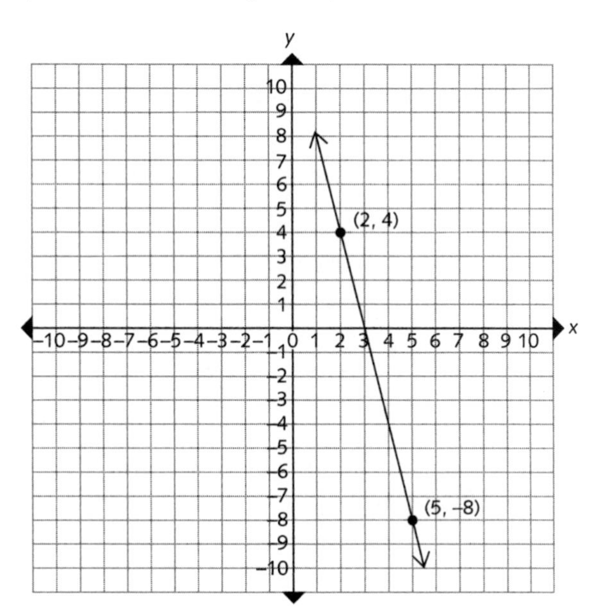

17

Graphing Linear Equations

In this chapter, you learn about graphing linear equations.

Properties of a Line

The graph of a linear equation is a straight line or simply a *line*. The line is the simplest graph of algebra but is probably the most referenced graph because it applies to many situations. Moreover, it has some unique properties that are exploited to great advantage in the study of mathematics. Here are two important properties of a line.

Two Important Properties of a Line

1. A line is completely determined by two distinct points.
2. Every nonvertical line has a unique number associated with it called the *slope*.

See Chapter 16 for an additional discussion of slope.

Graphing a Linear Equation That Is in Standard Form

The *standard form* of the equation of a line is $Ax + By = C$. For graphing purposes, the *slope-y-intercept form* $y = mx + b$ is the most useful. You use simple algebraic steps to put an equation of a line in this form.

If you have two distinct points (x_1, y_1) and (x_2, y_2) on a nonvertical line, then the slope, m, of the line is given by the ratio $m = \frac{y_2 - y_1}{x_2 - x_1}$. If the slope is negative, the line is slanting down as you move from left to right. If the slope is positive, the line is slanting up as you move from left to right. And if the slope is 0, the line is a horizontal line.

Problem Find the slope of the line that passes through the points (4,6) and (3,7).

Solution

Step 1. Use the slope ratio formula to find the slope.

$$m = \frac{y_2 - y_1}{x_2 - x_1} = \frac{6-7}{4-3} = \frac{-1}{1} = -1$$

When you graph linear equations "by hand," you put the equation in the slope-y-intercept form and then set up an x-y T-table (illustrated below) to compute point values for the graph.

Problem Graph the line whose equation is $y - 2x = -1$.

Solution

Step 1. Put the equation in slope-y-intercept form by solving for y.

$$y - 2x = -1$$
$$y - 2x + 2x = -1 + 2x$$
$$y = -1 + 2x$$
$$y = 2x - 1$$

Step 2. Set up an x-y T-table.

x	$y = 2x - 1$

Step 3. Because two points determine a line, substitute two convenient values for x and compute the corresponding y values.

x	$y = 2x - 1$
0	–1
1	1

Step 4. Plot the two points and draw the line through the points.

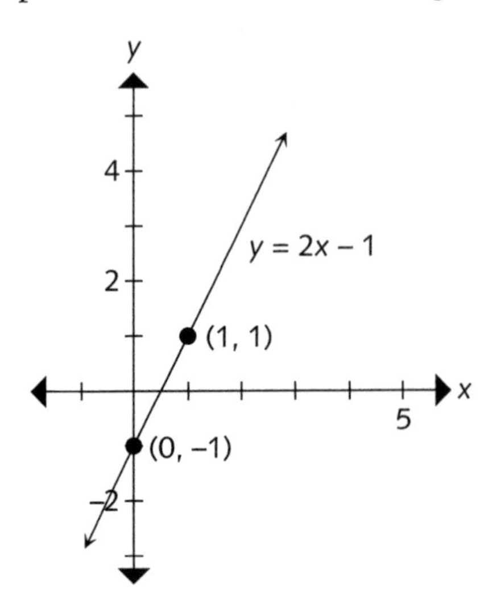

Graphing a Linear Equation That Is in Slope-*y*-Intercept Form

When the equation is in slope-y-intercept form, $y = mx + b$, the number m is the slope, and the number b is the y value of the point on the line where it crosses the y-axis. Hence, b is the y-intercept. The x value for the intersection point is always $x = 0$. In that case, you actually need to calculate only one y value to draw the line.

Problem Graph the line whose equation is $y = 3x + 1$.

Solution

Step 1. Substitute a value for x other than 0 and compute the corresponding y value.

When $x = 1$, $y = 3(1) + 1 = 4$.

Step 2. Plot the intercept point and the point $(1,4)$ and draw the graph.

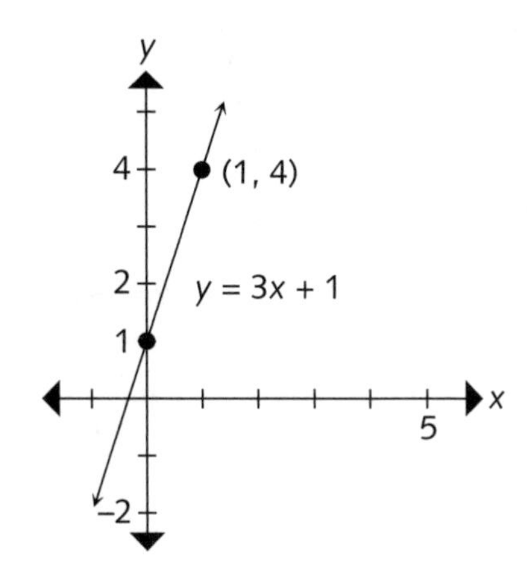

Observe from the graph that the run is 1 and the rise is 3, so the slope is $\frac{\text{rise}}{\text{run}} = \frac{3}{1} = 3$ as verified by the equation $y = 3x + 1$.

Problem Graph the line whose equation is $3x + 2y = 4$.

Solution

Step 1. Solve the equation for y to get the slope-y-intercept form.

$$3x + 2y = 4$$
$$3x + 2y - 3x = 4 - 3x$$
$$2y = -3x + 4$$
$$\frac{2y}{2} = \frac{-3x + 4}{2}$$
$$y = -\frac{3}{2}x + 2$$

Step 2. Choose a convenient x value, say, $x = 2$, and compute the corresponding y value.

$$y = \frac{-3}{2}(2) + 2 = -1.$$

Step 3. Plot the intercept point and the point $(2,-1)$ and draw the graph.

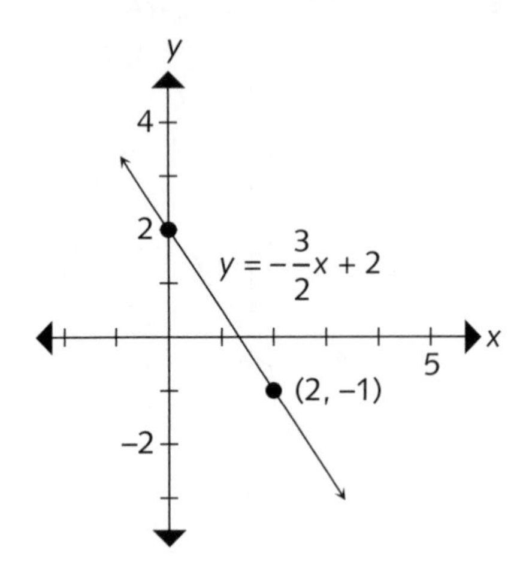

Observe from the graph that the run is -2 and the rise is 3, so the slope is $\frac{\text{rise}}{\text{run}} = \frac{3}{-2} = -\frac{3}{2}$ as verified by the equation $y = -\frac{3}{2}x + 2$.

As you can see, graphing linear equations is relatively simple because all you need is two points that lie on the graph of the equation. Finally, if you are using a graphing calculator to graph linear equations, the equation *must* be in slope-y-intercept form.

Exercise 17

1. Find the slope of the line through the points (5,11) and (3,14).

For 2–6, draw the graph of the line determined by the given equation.

2. $y = -2x + 6$
3. $3y = 5x - 9$
4. $y = x$
5. $4y - 5x = 8$
6. $y = \frac{1}{3}x - \frac{2}{3}$

18

The Equation of a Line

In this chapter, you determine the equation of a line. The basic graph of all of mathematics is the straight line. It is the simplest to draw, and it has the unique property that it is completely determined by just two distinct points. Because of this unique property, it is a simple matter to write the equation of a line given just two items of critical information.

There are three common methods for determining the equation of a line.

Determining the Equation of a Line Given the Slope and *y*-Intercept

This is the simplest of the methods for determining the equation of a line. You merely use the slope-y-intercept form of the equation of a line: $y = mx + b$.

Problem Given the slope $m = 3$ and the y-intercept $y = 5$, write the equation of the line.

Solution

Step 1. Recalling that the slope-y-intercept form of the equation of a line is $y = mx + b$, write the equation.

The equation of the line is $y = 3x + 5$. (You can see why this is the simplest method!)

Problem Given the slope $m = \frac{1}{2}$ and the y-intercept $y = -2$, write the equation of the line.

Solution

Step 1. Recalling that the slope-y-intercept form of the equation of a line is $y = mx + b$, write the equation.

The equation of the line is $y = \frac{1}{2}x - 2$.

Determining the Equation of a Line Given the Slope and One Point on the Line

For this method, you use the point-slope equation $m = \frac{y_1 - y_2}{x_1 - x_2}$, where (x_1, y_1) and (x_2, y_2) are points on the line.

Watch your signs when you use the point-slope equation.

Problem Given the slope $m = 2$ and a point (3, 2) on the line, write the equation of the line.

Solution

Step 1. Let (x, y) be a point on the line different from (3, 2), then substitute the given information into the point-slope formula: $m = \frac{y_1 - y_2}{x_1 - x_2}$.

$$2 = \frac{y - 2}{x - 3}$$

Step 2. Solve the equation for y to get the slope-y-intercept form of the equation.

$$2 = \frac{y - 2}{x - 3}$$

$$2(x - 3) = y - 2$$

$$2x - 6 = y - 2$$

$$2x - 6 + 2 = y - 2 + 2$$

$$2x - 4 = y$$

$y = 2x - 4$ is the equation of the line.

Problem Given the slope $m = \frac{1}{2}$ and a point (–1, 3) on the line, write the equation of the line.

Solution

Step 1. Let (x, y) be a point on the line different from (–1, 3), then substitute the given information into the point-slope formula: $m = \frac{y_1 - y_2}{x_1 - x_2}$.

$$\frac{y-3}{x-(-1)} = \frac{1}{2}$$

Step 2. Solve the equation for y to get the slope-y-intercept form of the equation.

$$\frac{y-3}{x+1} = \frac{1}{2}$$

$$y-3 = \frac{1}{2}(x+1)$$

$$y-3 = \frac{1}{2}x + \frac{1}{2}$$

$$y-3+3 = \frac{1}{2}x + \frac{1}{2} + 3$$

$y = \frac{1}{2}x + \frac{7}{2}$ is the equation of the line.

Problem Given the slope $m = -2$ and a point (0, 0) on the line, write the equation of the line.

Solution

Step 1. Let (x, y) be a point on the line different from (0, 0), then substitute the given information into the point-slope formula: $m = \frac{y_1 - y_2}{x_1 - x_2}$.

$$\frac{y-0}{x-0} = -2$$

Step 2. Solve the equation for y to get the slope-y-intercept form of the equation.

$$\frac{y-0}{x-0} = -2$$

$$\frac{y}{x} = -2$$

$y = -2x$ is the equation of the line.

Determining the Equation of a Line Given Two Distinct Points on the Line

You also use the point-slope equation with this method.

Problem Given the points (3, 4) and (1, 2) on the line, write the equation of the line.

Solution

Step 1. Use the two points to determine the slope using the point-slope equation.

$$m = \frac{4-2}{3-1} = \frac{2}{2} = 1$$

Step 2. Now use the point-slope formula and one of the given points to finish writing the equation. Let (x, y) be a point on the line different from, say, (3, 4).

$$\frac{y-4}{x-3} = 1$$

Step 3. Solve the equation for y to get the slope-y-intercept form of the equation.

$$\frac{y-4}{x-3} = 1$$

$y - 4 = x - 3$

$y - 4 + 4 = x - 3 + 4$

$y = x + 1$ is the equation of the line.

Problem Given the points (–1, 4) and (3, –7) on the line, write the equation of the line.

Solution

Step 1. Use the two points to determine the slope using the point-slope equation.

$$m = \frac{4-(-7)}{(-1)-3} = \frac{4+7}{-4} = -\frac{11}{4}$$

Step 2. Now use the point-slope formula and one of the given points to finish writing the equation. Let (x, y) be a point on the line different from, say, (3, –7).

$$\frac{y-(-7)}{x-3}=-\frac{11}{4}$$

Step 3. Solve the equation for y to get the slope-y-intercept form of the equation.

$$\frac{y-(-7)}{x-3}=-\frac{11}{4}$$

$$y+7=-\frac{11}{4}(x-3)$$

$$y+7=-\frac{11}{4}x+\frac{33}{4}$$

$$y+7-7=-\frac{11}{4}x+\frac{33}{4}-7$$

$$y=-\frac{11}{4}x+\frac{33}{4}-\frac{28}{4}$$

$y=-\frac{11}{4}x+\frac{5}{4}$ is the equation of the line.

When two points are known, it does not make any difference which one is chosen to finish writing the equation.

Exercise 18

1. Given the slope $m = 4$ and the y-intercept $y = 3$, write the equation of the line.
2. Given the slope $m = -3$ and the y-intercept $y = -3$, write the equation of the line.
3. Given the slope $m=\frac{1}{3}$ and the y-intercept $y = 0$, write the equation of the line.
4. Given the slope $m = 2$ and a point (1, 1) on the line, write the equation of the line.
5. Given the slope $m = -1$ and a point (2, 3) on the line, write the equation of the line.

6. Given the slope $m = \frac{1}{5}$ and a point (0, 1) on the line, write the equation of the line.
7. Given the points (2, 4) and (1, 2) on the line, write the equation of the line.
8. Given the points (–1, 2) and (1, 2) on the line, write the equation of the line.
9. Given the points (2, –1) and (1, 0) on the line, write the equation of the line.
10. Given the points (4, 4) and (6, 6) on the line, write the equation of the line.

19

Basic Function Concepts

Representations of a Function

Basic function concepts are presented in this chapter. One of the fundamental concepts of mathematics is the notion of a function. In algebra, you will see this idea in various settings, and you should become familiar with the different representations (forms) of a function. Three of the most frequently used representations are presented here.

Form 1: Ordered Pairs

A *function* is a set of ordered pairs such that no two different ordered pairs have the same *first* coordinate. The *domain* of a function is the set of all first coordinates of the ordered pairs in the function. The *range* of a function is the set of all second coordinates of the ordered pairs in the function.

Problem Determine which of the two sets is a function and identify the domain and range of the function.

$f = \{(4, 1), (3, 7), (2, 5), (5, 5)\}$ $\qquad w = \{(5, 1), (4, 3), (6, 3), (4, 2)\}$

Solution

Step 1. Analyze the sets for ordered pairs that satisfy the function criteria. Set f is a function because no ordered pairs have the same first coordinate. Set w is not a function because (4, 3) and (4, 2) have the same first coordinate, but different second coordinates.

Step 2. Isolate the first and second coordinates of the function f.

The domain of f is $D = \{2, 3, 4, 5\}$ and the range of f is $R = \{1, 5, 7\}$.

Suggestion: When listing the elements of the domain and range of a function, put the elements in numerical order.

Problem Identify the domain and range of the function $g = \{(1, 2), (2, 3), (3, 4), (4, 5), \ldots\}$.

Solution

Step 1. Isolate the first and second coordinates of g.

The domain of g is $D = \{1, 2, 3, 4, \ldots\}$, and the range of g is $R = \{2, 3, 4, 5, \ldots\}$.

Form 2: Equation or Rule

The ordered pair form is very useful for getting across the basic idea, but other forms are more useful for algebraic work.

A *function* is a *rule* of correspondence between two sets A and B such that each element in set A is paired with exactly one element in set B. In algebra, the *rule* is normally an equation in two variables. An example is the equation $y = 3x + 7$. For this *rule*, 1 is paired with 10, 2 with 13, and 5 with 22. This is equivalent to saying that the ordered pairs (1, 10), (2, 13) and (5, 22) are in the function.

If the domain of a function is not obvious (as it is in the first two problems) or not specified, then it is generally assumed that the domain is the largest set of real numbers for which the equation has numerical meaning in the set of real numbers. The domain, then, unless otherwise stated, is all the real numbers except excluded values. To determine the domain, start with the real numbers and exclude all values for x, if any, that would make the equation undefined over the real numbers.

Routinely, division by 0 and even roots of negative numbers create domain problems.

Problem State the domain of the given function.

a. $y = \dfrac{3}{x-1}$

b. $y = \sqrt{x-2} + 5$

Solution

a. $y = \frac{3}{x-1}$

Step 1. Set the denominator equal to 0 and solve for x.

$x - 1 = 0$

$x - 1 + 1 = 0 + 1$

$x = 1$

Step 2. State the domain.

The domain of $y = \frac{3}{x-1}$ is all real numbers except 1.

b. $y = \sqrt{x-2} + 5$

Step 1. Because the square root is an even root, set the term under the radical greater than or equal to 0 and solve the inequality.

$x - 2 \geq 0$

$x - 2 + 2 \geq 0 + 2$

$x \geq 2$

Step 2. State the domain.

The domain of $y = \sqrt{x-2} + 5$ is all real numbers greater than or equal to 2.

Terminology of Functions

A function is completely determined when the domain is known and the rule is specified. Even though the range is determined, it is often difficult to exhibit or specify the set of numbers in the range. Some of the techniques for determining the range of a function are beyond the scope of this book, and the focus here will be mostly on the domain and the equation that gives the *rule*. In fact, the words *rule* and *equation* will be used synonymously.

Some common terminology used in the study of functions is the following: (1) Domain values are called *inputs*, and range values are called *outputs*. (2) In equations of the form $y = 3x + 5$, x is called the *independent variable*, and y is called the *dependent variable*. Also, a convenient notation for a function is to use the symbol $f(x)$ to denote the value of the function f at a given value for x. In this setting, it is convenient to think of x as being an input value and $f(x)$ as being an output value. You can also write the function $y = 3x + 5$ as $f(x) = 3x + 5$, where $y = f(x)$.

The notation $f(x)$ does not mean f times x. It is a special notation for the value of a function.

Problem Find the value of the function $f(x) = 3x + 5$ at the given x value.

a. $x = 3$

b. $x = 0$

Solution

a. $x = 3$

Step 1. Check whether 3 is in the domain of the function.

The equation will generate real number values for each real number x. The domain, then, is all real numbers. Thus, 3 is in the domain of f.

Step 2. Substitute the given number for x in the equation and evaluate.

$f(3) = 3(3) + 5$

$f(3) = 14$

b. $x = 0$

Step 1. Check whether 0 is in the domain of the function.

The equation will generate real number values for each real number x. The domain, then, is all real numbers. Thus, 0 is in the domain of f.

Step 2. Substitute the given number for x in the equation and evaluate.

$f(0) = 3(0) + 5$

$f(0) = 5$

Problem Find the values of the function $g(x) = \sqrt{x-3} + 2x$ at the given x value.

a. $x = 4$

b. $x = 1$

c. $x = 8$

Solution

a. $x = 4$

Step 1. Check whether 4 is in the domain of the function.

The square root of a negative number is not a real number, so set $x - 3 \geq 0$ and solve the inequality to determine the domain of the function.

$$x - 3 \geq 0$$

$$x \geq 3$$

The domain is all real numbers greater than or equal to 3. Thus, 4 is in the domain of g.

Step 2. Substitute 4 for x in the equation and evaluate.

$$g(4) = \sqrt{4-3} + 2(4)$$

$$g(4) = \sqrt{1} + 8$$

$$g(4) = 9$$

b. $x = 1$

Step 1. Check whether 1 is in the domain of the function.

The square root of a negative number is not a real number, so set $x - 3 \geq 0$ and solve the inequality to determine the domain of the function.

$$x - 3 \geq 0$$

$$x \geq 3$$

The domain is all real numbers greater than or equal to 3. Thus, 1 is not in the domain of g, so the function has no value at 1.

c. $x = 8$

Step 1. Check whether 8 is in the domain of the function.

The square root of a negative number is not a real number, so set $x - 3 \geq 0$ and solve the inequality to determine the domain of the function.

$$x - 3 \geq 0$$

$$x \geq 3$$

The domain is all real numbers greater than or equal to 3. Thus, 8 is in the domain of g.

Step 2. Substitute 8 for x in the equation and evaluate.

$$g(8) = \sqrt{8-3} + 2(8)$$

$$g(8) = \sqrt{5} + 16$$

Problem Find the value of the function $f(x) = \sqrt[3]{x+1} + 3$ at the given x value.

a. $x = -9$
b. $x = 0$

Solution

a. $x = -9$

Step 1. Check whether -9 is in the domain of the function.

Because the cube root is an odd root number, there is no restriction on the values under the radical. The domain, then, is all real numbers, so -9 is in the domain of f.

Step 2. Substitute -9 for x in the equation and evaluate.

$$f(-9) = \sqrt[3]{(-9)+1} + 3$$

$$f(-9) = \sqrt[3]{-8} + 3$$

$$f(-9) = -2 + 3 = 1$$

b. $x = 0$

Step 1. Check whether 0 is in the domain of the function.

Because the cube root is defined for all real numbers, there is no restriction on the values under the radical. The domain, then, is all real numbers, so 0 is in the domain of f.

Step 2. Substitute 0 for x in the equation and evaluate.

$$f(0) = \sqrt[3]{(0)+1} + 3$$

$$f(0) = \sqrt[3]{1} + 3 = 4$$

Form 3: Graphical Representation

An additional way to represent a function is by graphing the function in the Cartesian coordinate plane. You can easily determine whether a graph is the graph of a function by using the *vertical line test*. A graph is the graph of a function if and only if no vertical line crosses the graph in more than one point. This is a quick visual determination and is the graphical equivalent of saying that no two different ordered pairs have the same first coordinate.

Problem Determine which of the following is the graph of a function.

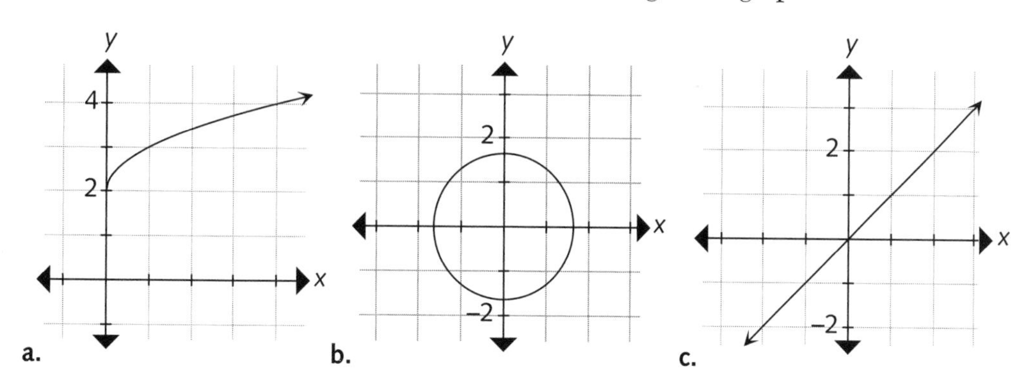

Solution

Step 1. Mentally apply the vertical line test to each graph.

a. This is the graph of a function.

b. This is not the graph of a function because a vertical line drawn through the point $x = 1$ will cross the graph in more than one point. (Actually, there are infinitely many vertical lines that will cross in more than one point, but it only takes one to ascertain it is not a function.)

c. This is the graph of a function.

Some Common Functions

Some of the more common functions you will study in algebra are listed here in general form and given special names.

a. $y = f(x) = ax + b$ Linear function

b. $y = f(x) = ax^2 + bx + c,\ a \neq 0$ Quadratic function

c. $y = |x|$ Absolute value function

d. $y = \sqrt{x}$ Square root function

Chapter 17 dealt with the graph of linear functions. Sample graphs of the four functions above are shown in Figure 19.1. These are easily graphed with a graphing calculator, which is a good tool to have when you study algebra.

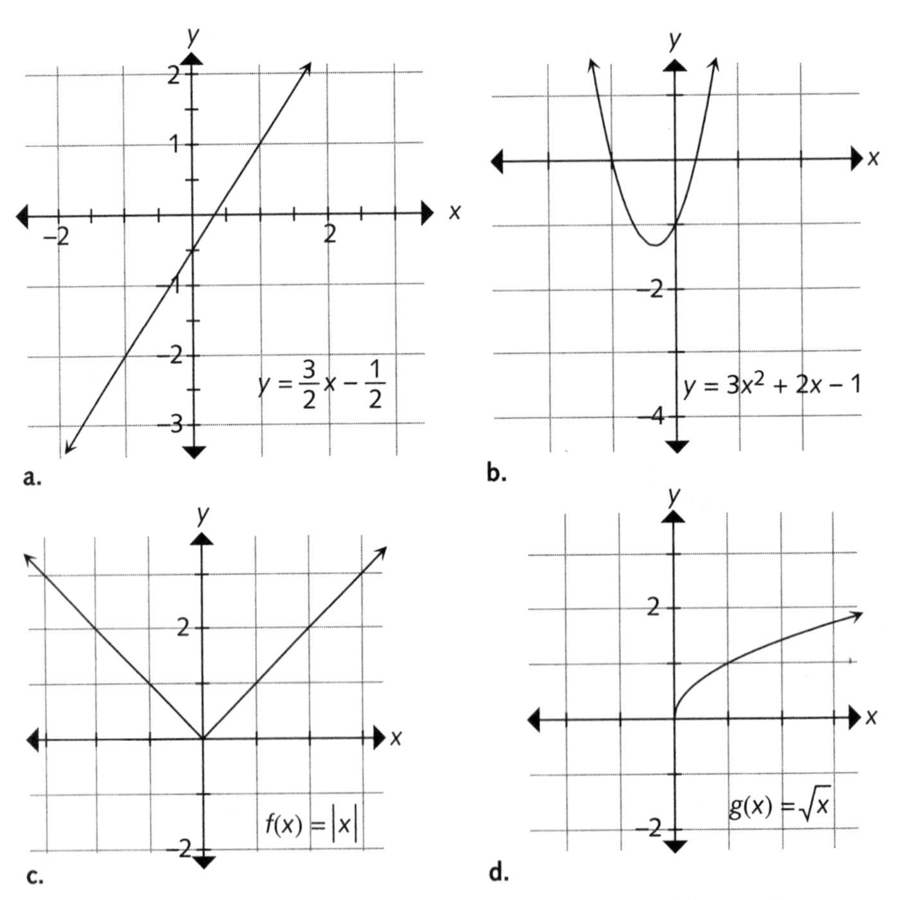

Figure 19.1 Sample graphs of four common functions: a) linear function; b) quadratic function; c) absolute value function; d) square root function

Of course, you can graph these functions "by hand" by setting up an x-y T-table and substituting several representative values for x.

Functional relationships naturally occur in many and various circumstances. A few examples will illustrate.

Problem Establish a functional relationship between the radius of a circle and its area.

Solution

Step 1. The formula for the area of a circle is πr^2, where r is the radius.

Area = $A = \pi r^2$

Step 2. Express the area of a circle as a function of its radius.

$A(r) = \pi r^2$

Problem Establish a functional relationship between the x and y values in the following table.

x	3	4	5	6	7
y	7	9	11	13	15

Solution

Step 1. Look for a pattern that will connect the two numbers and describe the pattern in words.

The y number is twice the x number plus 1.

Step 2. Write the pattern for y in terms of x.

$y = 2x + 1$

Exercise 19

1. Determine which of the sets are functions.

   a. $f = \{(2, 1), (4, 5), (6, 9), (5, 9)\}$

   b. $g = \{(3, 4), (5, 1), (6, 3), (3, 6)\}$

   c. $h = \{(2, 1)\}$

   d. $t = \{(7, 5), (8, 9), (8, 9)\}$

2. State the domain and range of the function $g = \{(4, 5), (8, 9), (7, 7), (6, 7)\}$.

3. Find the domain of each of the functions.

   a. $y = f(x) = 5x - 7$

   b. $y = g(x) = \sqrt{2x - 3} + 4$

   c. $y = \dfrac{9x + 1}{x - 5}$

   d. $y = \dfrac{2x^2 + 5}{x^2 - 4}$

4. If $f(x) = 5\sqrt{x + 2} - 3$, find the indicated value.

   a. $f(2)$

   b. $f(-1)$

   c. $f(6)$

   d. $f(-3)$

5. Which of the graphs is the graph of a function?

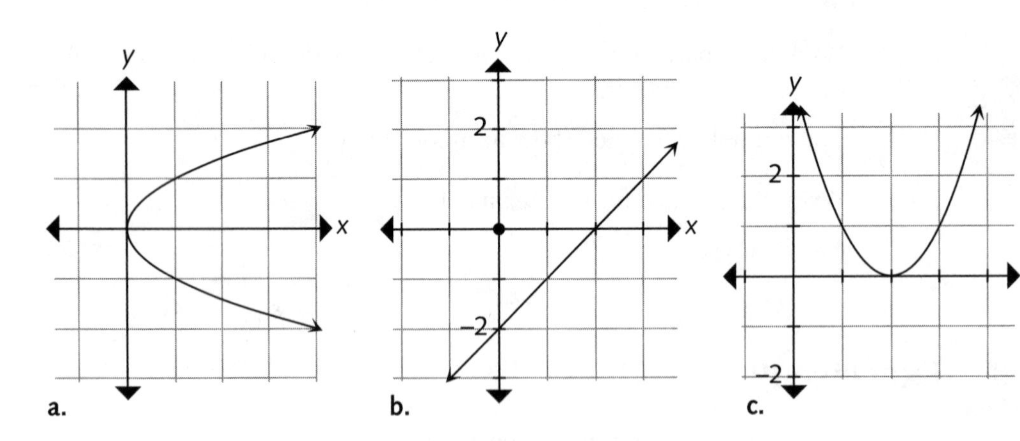

6. Write the equation for the functional relationship between x and y.

x	2	3	4	5
y	9	13	17	21

20

Systems of Equations

In algebra, you might need to solve two linear equations in two variables and to solve them simultaneously. Three methods for solving two simultaneous equations are presented in this chapter. Each has its strong and weak points, as will be pointed out in the problems.

Solutions to a System of Equations

From Chapter 19, the equation of a linear function has several forms, but for the purposes of this chapter, the form $ax + by = c$ is preferred. You know from Chapter 17 that the graph of a linear equation is a line. When you encounter two such equations, the basic question you should ask yourself is, "What are the coordinates of the point of intersection, if any, of the two lines?"

> Remember from Chapter 16 that the location of a point is an ordered pair of coordinates such as (x, y).

This question has three possible answers. If the lines do not intersect, there is no solution. If the two lines intersect, there is only one solution—an ordered pair (x, y). If the two lines are equal versions of the same line, then there are infinitely many solutions—an infinite set of ordered pairs.

Finally, when you are solving two linear equations in two variables, an example of the standard form of writing them together is

$$\begin{aligned} 2x - y &= 0 \\ x + y &= 3 \end{aligned}$$

You *solve the system* when you answer the question: What are the coordinates of the point of intersection, if any, of the two lines? Here are three methods for solving a system of equations.

Solving a System of Equations by Substitution

To solve a system of equations by substitution, you solve one equation for one of the variables in terms of the other variable and then use substitution to solve the system. (See Chapter 14 for a discussion on how to solve linear equations.)

Problem Solve the system.

$$2x - y = 0$$
$$x + y = 3$$

Solution

Step 1. Solve the first equation, $2x - y = 0$, for y in terms of x.

$2x - y = 0$

$2x - y + y = 0 + y$

$2x = y$

Step 2. Substitute $2x$ for y in the second equation, $x + y = 3$, and solve for x.

When you use the substitution method, enclose substituted values in parentheses to avoid errors.

$x + (\mathbf{2x}) = 3$

$3x = 3$

$\frac{3x}{3} = \frac{3}{3}$

$x = 1$

When you use the substitution method, you can substitute the value for x in either equation. Just pick the one you think would be easier to work with.

Step 3. Substitute 1 for x in the second equation, $x + y = 3$, and solve for y.

$\mathbf{1} + y = 3$

$1 + y - 1 = 3 - 1$

$y = 2$

Always check your solution in both equations when you solve a system of two linear equations.

Step 4. Check whether $x = 1$ and $y = 2$ satisfy both equations in the system.

$$2x - y = 0 \qquad 2(\mathbf{1}) - (\mathbf{2}) = 2 - 2 = 0$$
$$x + y = 3 \qquad (\mathbf{1}) + (\mathbf{2}) = 1 + 2 = 3$$
Check. √

Step 5. Write the solution.

The solution is $x = 1$ and $y = 2$. That is, the two lines intersect at the point (1, 2).

When you use the substitution method, it makes no difference which equation is solved first or for which variable, but when you solve it, be sure to substitute the value in the other equation.

Problem Solve the system.

$$2x - y = 4$$
$$x + y = 5$$

Solution

Step 1. Solve the second equation, $x + y = 5$, for x in terms of y.

$x + y = 5$

$x + y - y = 5 - y$

$x = 5 - y$

Step 2. Substitute $5 - y$ for x in the first equation, $2x - y = 4$, and solve for y.

$2(\mathbf{5 - y}) - y = 4$

$10 - 2y - y = 4$

$10 - 3y = 4$

$10 - 3y - 10 = 4 - 10$

$-3y = -6$

$\frac{-3y}{-3} = \frac{-6}{-3}$

$y = 2$

Step 3. Substitute 2 for y in the second equation, $x + y = 5$, and solve for x.

$x + (\mathbf{2}) = 5$

$x + 2 = 5$

$x + 2 - 2 = 5 - 2$

$x = 3$

Step 4. Check whether $x = 3$ and $y = 2$ satisfy both equations.

$$2x - y = 4 \qquad 2(\mathbf{3}) - (\mathbf{2}) = 6 - 2 = 4$$
$$x + y = 5 \qquad (\mathbf{3}) + (\mathbf{2}) = 3 + 2 = 5$$

Check. √

Step 5. Write the solution.

The solution is $x = 3$ and $y = 2$. That is, the two lines intersect at the point (3, 2).

Solving a System of Equations by Elimination

To solve a system of equations by elimination, you multiply the equations by constants to produce opposite coefficients of one variable so that it can be eliminated by adding the two equations.

Problem Solve the system.

$$2x - y = 4$$
$$x + 2y = -3$$

Solution

Step 1. To eliminate x, multiply the second equation by -2.

$$2x - y = 4 \longrightarrow 2x - y = 4$$
$$x + 2y = -3 \xrightarrow[\text{Multiply by } -2]{} -2x - 4y = 6$$

Step 2. Add the resulting two equations.

$$2x - y = 4$$
$$\underline{-2x - 4y = 6}$$
$$-5y = 10$$

Step 3. Solve $-5y = 10$ for y.

$$-5y = 10$$
$$\frac{-5y}{-5} = \frac{10}{-5}$$
$$y = -2$$

Step 4. Substitute -2 for y in one of the original equations, $2x - y = 4$, and solve for x.

$$2x - (\mathbf{-2}) = 4$$

$$2x + 2 = 4$$
$$2x = 2$$
$$\frac{2x}{2} = \frac{2}{2}$$
$$x = 1$$

Step 5. Check whether $x = 1$ and $y = -2$ satisfy both original equations.

$$\begin{array}{ll} 2x - y = 4 & 2(\mathbf{1}) - (\mathbf{-2}) = 2 + 2 = 4 \\ x + 2y = -3 & (\mathbf{1}) + 2(\mathbf{-2}) = 1 - 4 = -3 \end{array} \text{ Check. } \surd$$

Step 6. Write the solution.

The solution is $x = 1$ and $y = -2$. That is, the two lines intersect at the point $(1, -2)$.

Problem Solve the system.

$$5x + 2y = 3$$
$$2x + 3y = -1$$

Solution

Step 1. To eliminate y, multiply the first equation by 3 and the second equation by -2.

$$5x + 2y = 3 \xrightarrow{\text{Multiply by 3}} 15x + 6y = 9$$
$$2x + 3y = -1 \xrightarrow[\text{Multiply by } -2]{} -4x - 6y = 2$$

Step 2. Add the resulting two equations.

$$\begin{array}{r} 15x + 6y = 9 \\ \underline{-4x - 6y = 2} \\ 11x = 11 \end{array}$$

Step 3. Solve $11x = 11$ for x.

$$11x = 11$$
$$\frac{11x}{11} = \frac{11}{11}$$
$$x = 1$$

Step 4. Substitute 1 for x in one of the original equations, $5x + 2y = 3$, and solve for y.

$5(\mathbf{1}) + 2y = 3$

$5 + 2y = 3$

$5 + 2y - 5 = 3 - 5$

$2y = -2$

$\frac{2y}{2} = \frac{-2}{2}$

$y = -1$

Step 5. Check whether $x = 1$ and $y = -1$ satisfy both original equations.

$5x + 2y = 3 \quad 5(\mathbf{1}) + 2(\mathbf{-1}) = 5 - 2 = 3$

$2x + 3y = -1 \quad 2(\mathbf{1}) + 3(\mathbf{-1}) = 2 - 3 = -1$

Check. √

Step 6. Write the solution.

The solution is $x = 1$ and $y = -1$. That is, the two lines intersect at the point $(1, -1)$.

Solving a System of Equations by Graphing

To solve a system of equations by graphing, you graph the two equations and locate (as accurately as possible) the intersection point on the graph. Because graphing devices such as graphing calculators and computer algebra systems require the slope-intercept form of the equation of straight lines, the steps will include writing the equations in that form. (See Chapter 17 for a discussion of the slope-intercept form.)

> The graphing method might yield inaccurate results due to the limitations of graphing.

Problem Solve the system.

$2x + 5y = 3$

$3x - 2y = 1$

Solution

Step 1. Write both equations in slope-intercept form.

The first equation, $2x + 5y = 3$, yields $y = -\frac{2}{5}x + \frac{3}{5}$.

The second equation, $3x - 2y = 1$, yields $y = \frac{3}{2}x - \frac{1}{2}$.

Step 2. Graph both equations on the same graph.

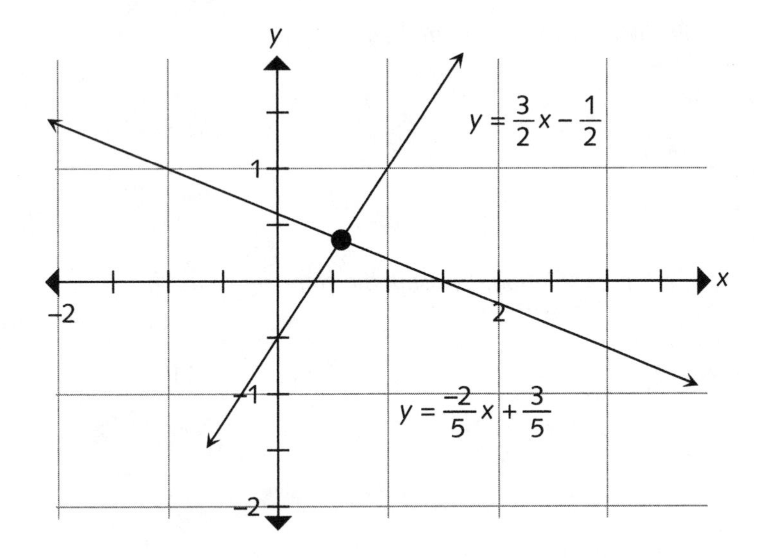

You can find an approximate solution of $x \approx 0.579$ and $y \approx 0.368$ by using a graphing utility. These values are close estimates but will not exactly satisfy either equation. Of course, you can find the exact solution of $x = \frac{11}{19}$ and $y = \frac{7}{19}$ by using either the substitution method or the elimination method. Nevertheless, the graphical approach gives you the approximate location of the intercept (if any), and, more important, this method helps you see the connection between the solution and the graphs of the two equations.

When an exact solution is needed, do *not* use the graphing method for solving a system of equations.

Exercise 20

For 1–3, solve by the substitution method.

1. $x - 2y = -4$
   $2x + y = 7$
2. $4x - y = 3$
   $x - 3y = -13$
3. $4x + 2y = 8$
   $2x - 3y = -8$

For 4–6, solve by the elimination method.

4. $-2x + 4y = 8$
   $-2x - y = -7$
5. $8x - 2y = 6$
   $x - 3y = -13$
6. $2x + y = 4$
   $4x - 6y = -16$

For 7 and 8, estimate solutions by using the graphing method.

7. $3x - 2y = 3$
   $6x + 2y = 9$
8. $7x + 14y = 2$
   $14x - 7y = -11$

Answer Key

Chapter 1 Numbers of Algebra

Exercise 1

1. 10 is a natural number, a whole number, an integer, a rational number, and a real number.
2. $\sqrt{0.64} = 0.8$ is a rational number and a real number.
3. $\sqrt[3]{\frac{8}{125}} = \frac{2}{5}$ is a rational number and a real number.
4. $-\pi$ is an irrational number and a real number.
5. -1000 is an integer, a rational number, and a real number.
6. $\sqrt{2}$ is an irrational number and a real number.
7. $-\sqrt{\frac{3}{4}}$ is an irrational number and a real number.
8. $-\sqrt{\frac{9}{4}} = -\frac{3}{2}$ is a rational number and a real number.
9. 1 is a natural number, a whole number, an integer, a rational number, and a real number.
10. $\sqrt[3]{0.001} = 0.1$ is a rational number and a real number.
11. Closure property of multiplication
12. Commutative property of addition
13. Multiplicative inverse property

14. Closure property of addition
15. Associative property of addition
16. Distributive property
17. Additive inverse property
18. Zero factor property
19. Associative property of multiplication
20. Multiplicative identity property

Chapter 2 Computation with Real Numbers

Exercise 2

1. $|-45| = 45$
2. $|5.8| = 5.8$
3. $\left|-5\frac{2}{3}\right| = 5\frac{2}{3}$
4. "Negative nine plus the opposite of negative four equals negative nine plus four"
5. "Negative nine minus negative four equals negative nine plus four"
6. $-80 + -40 = -120$
7. $0.7 + -1.4 = -0.7$
8. $\left(-\frac{5}{6}\right)\left(\frac{2}{5}\right) = -\frac{10}{30} = -\frac{1}{3}$
9. $\frac{18}{-3} = -6$
10. $(-100)(-8) = 800$
11. $(400)\left(\frac{1}{2}\right) = 200$
12. $\frac{-1\frac{1}{3}}{-\frac{1}{3}} = \frac{-\frac{4}{3}\cdot 3}{-\frac{1}{3}\cdot 3} = \frac{-4}{-1} = 4$
13. $-450.95 - (-65.83) = -385.12$
14. $\frac{3}{11} - \left(-\frac{5}{11}\right) = \frac{3}{11} + \frac{5}{11} = \frac{8}{11}$

15. $\frac{0.8}{-0.01} = -80$
16. $-458 + 0 = -458$
17. $\left(4\frac{1}{2}\right)\left(-3\frac{3}{5}\right)(0)(999)\left(-\frac{5}{17}\right)=0$
18. $\frac{0}{8.75}=0$
19. $\frac{700}{0}=\text{undefined}$
20. $(-3)(1)(-1)(-5)(-2)(2)(-10) = -600$

Chapter 3 Roots and Radicals

Exercise 3

1. 12 and –12
2. $\frac{5}{7}$ and $-\frac{5}{7}$
3. 0.8 and –0.8
4. 20 and –20
5. $\sqrt{16}=4$
6. $\sqrt{-9}$ not a real number
7. $\sqrt{\frac{16}{25}}=\frac{4}{5}$
8. $\sqrt{25+144}=\sqrt{169}=13$
9. $\sqrt{-5\cdot-5}=|-5|=5$
10. $\sqrt{z\cdot z}=|z|$
11. $\sqrt[3]{-125}=-5$
12. $\sqrt[3]{\frac{64}{125}}=\frac{4}{5}$
13. $\sqrt[3]{0.027}=0.3$
14. $\sqrt[3]{y\cdot y\cdot y}=y$
15. $\sqrt[4]{625}=5$
16. $\sqrt[5]{-\frac{32}{243}}=-\frac{2}{3}$
17. $\sqrt[6]{-64}$ not a real number
18. $\sqrt[7]{0}=0$
19. $\sqrt{72}=\sqrt{36\cdot 2}=\sqrt{36}\cdot\sqrt{2}=6\sqrt{2}$
20. $\sqrt{\frac{2}{3}}=\sqrt{\frac{2\cdot 3}{3\cdot 3}}=\sqrt{\frac{6}{9}}=\sqrt{\frac{1}{9}\cdot 6}=\sqrt{\frac{1}{9}}\cdot\sqrt{6}=\frac{1}{3}\sqrt{6}$

Chapter 4 Exponentiation

Exercise 4

1. $-4\cdot-4\cdot-4\cdot-4\cdot-4=(-4)^5$
2. $8\cdot 8\cdot 8\cdot 8\cdot 8\cdot 8\cdot 8=8^7$
3. $(-2)^7=-128$
4. $(0.3)^4=0.0081$
5. $\left(-\frac{3}{4}\right)^2=\frac{9}{16}$
6. $0^9=0$
7. $(1+1)^5=2^5=32$

8. $(-2)^0 = 1$

9. $3^{-4} = \frac{1}{3^4} = \frac{1}{81}$

10. $(-4)^{-2} = \frac{1}{(-4)^2} = \frac{1}{16}$

11. $(0.3)^{-2} = \frac{1}{(0.3)^2} = \frac{1}{0.09}$

12. $\left(\frac{3}{4}\right)^{-1} = \frac{1}{(3/4)^1} = \frac{1}{3/4} = \frac{4}{3}$

13. $(-125)^{1/3} = \sqrt[3]{-125} = -5$

14. $(0.16)^{1/2} = \sqrt{0.16} = 0.4$

15. $(-121)^{1/4} = \sqrt[4]{-121}$ not a real number

16. $\left(\frac{16}{625}\right)^{1/4} = \sqrt[4]{\frac{16}{625}} = \frac{2}{5}$

17. $(-27)^{2/3} = \left[(-27)^{1/3}\right]^2 = [-3]^2 = 9$

18. $\left(\frac{16}{625}\right)^{3/4} = \left(\sqrt[4]{\frac{16}{625}}\right)^3 = \left(\frac{2}{5}\right)^3 = \frac{8}{125}$

19. $\frac{1}{5^{-3}} = \frac{5^3}{1} = \frac{125}{1} = 125$

20. $\frac{1}{(-2)^{-4}} = \frac{(-2)^4}{1} = \frac{16}{1} = 16$

Chapter 5 Order of Operations

Exercise 5

1. $(5 + 7)6 - 10$
   $= 12 \cdot 6 - 10$
   $= 72 - 10$
   $= 62$

2. $(-7^2)(6 - 8)$
   $= (-7^2)(-2)$
   $= (-49)(-2)$
   $= 98$

3. $(2 - 3)(-20)$
   $= (-1)(-20)$
   $= 20$

4. $3(-2) - \frac{10}{-5}$
   $= -6 - \frac{10}{-5}$
   $= -6 - (-2)$
   $= -6 + 2$
   $= -4$

5. $9 - \frac{20 + 22}{6} - 2^3$
   $= 9 - \frac{42}{6} - 2^3$
   $= 9 - \frac{42}{6} - 8$
   $= 9 - 7 - 8$
   $= -6$

6. $-2^2 \cdot -3 - (15 - 4)^2$
   $= -2^2 \cdot -3 - (11)^2$
   $= -4 \cdot -3 - 121$
   $= 12 - 121$
   $= -109$

7. $5(11 - 3 - 6 \cdot 2)^2$
   $= 5(11 - 3 - 12)^2$
   $= 5(-4)^2$
   $= 5(16)$
   $= 80$

8. $-10-\dfrac{-8-(3\cdot -3+15)}{2}$

$=-10-\dfrac{-8-(6)}{2}$

$=-10-\dfrac{-14}{2}$

$= -10 - -7$

$= -10 + 7$

$= -3$

9. $\dfrac{7^2-8\cdot 5+3^4}{3\cdot 2-36\div 12}$

$=\dfrac{49-8\cdot 5+81}{3\cdot 2-36\div 12}$

$=\dfrac{49-40+81}{6-3}$

$=\dfrac{90}{3}$

$= 30$

10. $(-6)\left(\dfrac{\sqrt{625-576}}{14}\right)+\dfrac{6}{-3}$

$=(-6)\left(\dfrac{\sqrt{49}}{14}\right)+\dfrac{6}{-3}$

$=(-6)\left(\dfrac{7}{14}\right)+\dfrac{6}{-3}$

$=(-6)\left(\dfrac{1}{2}\right)+\dfrac{6}{-3}$

$= -3 - 2$

$= -5$

11. $\dfrac{5-|-5|}{20^2}$

$=\dfrac{5-5}{400}$

$=\dfrac{0}{400}$

$= 0$

12. $(12 - 5) - (5 - 12)$

$= 7 - (-7)$

$= 7 + 7$

$= 14$

13. $\dfrac{9+\sqrt{100-64}}{-|-15|}$

$=\dfrac{9+\sqrt{36}}{-15}$

$=\dfrac{9+6}{-15}$

$=\dfrac{15}{-15}$

$= -1$

14. $-8 + 2(-1)^2 + 6$

$= -8 + 2 \cdot 1 + 6$

$= -8 + 2 + 6$

$= 0$

15. $\dfrac{3}{2}\left(-\dfrac{2}{3}\right)-\dfrac{1}{4}(-5)+\dfrac{15}{7}\left(-\dfrac{7}{3}\right)$

$=-1+\dfrac{5}{4}-5$

$= -6 + 1.25$

$= -4.75$

Chapter 6 Algebraic Expressions

Exercise 6

1. Name the variable(s) and constant(s) in the expression $2\pi r$, where r is the measure of the radius of a circle. Answer: The letter r stands for the measure of the radius

of a circle and can be any real nonzero number, so r is a variable. The numbers 2 and π have fixed, definite values, so they are constants.

2. −12 is the numerical coefficient
3. 1 is the numerical coefficient
4. $\frac{2}{3}$ is the numerical coefficient
5. $-5x = -5 \cdot 9 = -45$
6. $2xyz = 2(9)(-2)(-3) = 108$
7. $\frac{6(x+1)}{5\sqrt{x}-10} = \frac{6(9+1)}{5\sqrt{9}-10}$
$= \frac{6 \cdot 10}{5 \cdot 3 - 10}$
$= \frac{60}{15-10}$
$= \frac{60}{5}$
$= 12$
8. $\frac{-2|y|+5(2x-y)}{-6z+y^3} = \frac{-2|-2|+5(2 \cdot 9-(-2))}{-6(-3)+(-2)^3}$
$= \frac{-2 \cdot 2+5(18+2)}{-6(-3)-8}$
$= \frac{-2 \cdot 2+5 \cdot 20}{-6(-3)-8}$
$= \frac{-4+100}{18-8}$
$= \frac{96}{10}$
$= 9.6$
9. $x^2 - 8x - 9 = 9^2 - 8 \cdot 9 - 9$
$= 81 - 72 - 9$
$= 0$
10. $2y + x(y - z) = 2(-2) + 9((-2) - (-3))$
$= 2(-2) + 9(-2 + 3)$
$= 2(-2) + 9(1)$
$= -4 + 9$
$= 5$

11. $\dfrac{(x+y)^2}{x^2-y^2}=\dfrac{(9+(-2))^2}{9^2-(-2)^2}$

$=\dfrac{(7)^2}{9^2-(-2)^2}$

$=\dfrac{49}{81-4}$

$=\dfrac{49}{77}$

$=\dfrac{7}{11}$

12. $(y + z)^{-3} = ((-2)+(-3))^{-3}$

$= (-2 - 3)^{-3}$

$= (-5)^{-3}$

$=\dfrac{1}{(-5)^3}$

$=\dfrac{1}{-125}$

$=-\dfrac{1}{125}$

13. $A=\dfrac{1}{2}bh=\dfrac{1}{2}\cdot 12\cdot 8=48$

14. $V=\dfrac{1}{3}\pi r^2h=\dfrac{1}{3}(3.14)(5^2)(18)=471$

15. $c^2 = a^2 + b^2$

$c^2 = 8^2 + 15^2$

$c^2 = 64 + 225$

$c^2 = 289$

c is 17 or −17

16. $-\left(-\dfrac{1}{2}x^3y^2+7xy^3-30\right)=\dfrac{1}{2}x^3y^2-7xy^3+30$

17. $(8a^3 + 64b^3) = 8a^3 + 64b^3$

18. $-4 - (-2y^3) = -4 + 2y^3$

19. $-3(x + 4) = -3x - 12$

20. $12 + (x^2 + y) = 12 + x^2 + y$

Chapter 7 Rules for Exponents

Exercise 7

1. $x^4x^9 = x^{13}$
2. $x^3x^4y^6y^5 = x^7y^{11}$
3. $\frac{x^6}{x^3} = x^3$
4. $\frac{x^5y^5}{x^2y^4} = x^3y$
5. $\frac{x^4}{x^6} = x^{-2} = \frac{1}{x^2}$
6. $(x^2)^5 = x^{10}$
7. $(xy)^5 = x^5y^5$
8. $(-5x)^3 = -125x^3$
9. $(2x^5yz^3)^4 = 16x^{20}y^4z^{12}$
10. $\left(\frac{5}{3x}\right)^4 = \frac{625}{81x^4}$
11. $\left(\frac{-3x}{5y}\right)^4 = \frac{81x^4}{625y^4}$
12. $(2x + 1)^2$ is a power of a sum. It cannot be simplified using only rules for exponents.
13. $(3x - 5)^3$ is a power of a difference. It cannot be simplified using only rules for exponents.
14. $(x + 3)(x + 3)^2 = (x + 3)^3$
15. $\frac{(2x-y)^{15}}{(2x-y)^5} = (2x-y)^{10}$

Chapter 8 Adding and Subtracting Polynomials

Exercise 8

1. $x^2 - x + 1$ is a trinomial.
2. $125x^3 - 64y^3$ is a binomial.
3. $2x^2 + 7x - 4$ is a trinomial.
4. $-\frac{1}{3}x^5y^2$ is a monomial.
5. $2x^4 + 3x^3 - 7x^2 - x + 8$ is a polynomial.
6. $-15x + 17x = 2x$
7. $14xy^3 - 7x^3y^2$ is simplified.
8. $10x^2 - 2x^2 - 20x^2 = -12x^2$
9. $10 + 10x$ is simplified.
10. $12x^3 - 5x^2 + 10x - 60 + 3x^3 - 7x^2 - 1$
    $= 15x^3 - 12x^2 + 10x - 61$
11. $(10x^2 - 5x + 3) + (6x^2 + 5x - 13)$
    $= 10x^2 - 5x + 3 + 6x^2 + 5x - 13$
    $= 16x^2 - 10$
12. $(20x^3 - 3x^2 - 2x + 5) + (9x^3 + x^2 + 2x - 15)$
    $= 20x^3 - 3x^2 - 2x + 5 + 9x^3 + x^2 + 2x - 15$
    $= 29x^3 - 2x^2 - 10$
13. $(10x^2 - 5x + 3) - (6x^2 + 5x - 13)$
    $= 10x^2 - 5x + 3 - 6x^2 - 5x + 13$
    $= 4x^2 - 10x + 16$
14. $(20x^3 - 3x^2 - 2x + 5) - (9x^3 + x^2 + 2x - 15)$
    $= 20x^3 - 3x^2 - 2x + 5 - 9x^3 - x^2 - 2x + 15$
    $= 11x^3 - 4x^2 - 4x + 20$

Chapter 9 Multiplying Polynomials

Exercise 9

1. $(4x^5y^3)(-3x^2y^3) = -12x^7y^6$
2. $(-8a^4b^3)(5ab^2) = -40a^5b^5$
3. $(-10x^3)(-2x^2) = 20x^5$
4. $(-3x^2y^5)(6xy^4)(-2xy) = 36x^4y^{10}$
5. $3(x - 5) = 3x - 15$
6. $x(3x^2 - 4) = 3x^3 - 4x$
7. $-2a^2b^3(3a^2 - 5ab^2 - 10)$
   $= -2a^2b^3 \cdot 3a^2 + 2a^2b^3 \cdot 5ab^2 + 2a^2b^3 \cdot 10$
   $= -6a^4b^3 + 10a^3b^5 + 20a^2b^3$
8. $(2x - 3)(x + 4)$
   $= 2x^2 + 8x - 3x - 12$
   $= 2x^2 + 5x - 12$
9. $(x + 4)(x + 5)$
   $= x^2 + 5x + 4x + 20$
   $= x^2 + 9x + 20$
10. $(x - 4)(x - 5)$
    $= x^2 - 5x - 4x + 20$
    $= x^2 - 9x + 20$
11. $(x + 4)(x - 5)$
    $= x^2 - 5x + 4x - 20$
    $= x^2 - x - 20$
12. $(x - 4)(x + 5)$
    $= x^2 + 5x - 4x - 20$
    $= x^2 + x - 20$
13. $(x - 1)(2x^2 - 5x + 3)$
    $= 2x^3 - 5x^2 + 3x - 2x^2 + 5x - 3$
    $= 2x^3 - 7x^2 + 8x - 3$
14. $(2x^2 + x - 3)(5x^2 - x - 2)$
    $= 10x^4 - 2x^3 - 4x^2 + 5x^3 - x^2 - 2x - 15x^2 + 3x + 6$
    $= 10x^4 + 3x^3 - 20x^2 + x + 6$
15. $(x - y)^2$
    $= (x - y)(x - y)$
    $= x^2 - xy - xy + y^2$
    $= x^2 - 2xy + y^2$
16. $(x + y)(x - y)$
    $= x^2 - xy + xy - y^2$
    $= x^2 - y^2$
17. $(x + y)^3$
    $= (x + y)(x + y)(x + y)$
    $= (x + y)(x^2 + 2xy + y^2)$
    $= x^3 + 2x^2y + xy^2 + x^2y + 2xy^2 + y^3$
    $= x^3 + 3x^2y + 3xy^2 + y^3$
18. $(x - y)^3$
    $= (x - y)(x - y)(x - y)$
    $= (x - y)(x^2 - 2xy + y^2)$
    $= x^3 - 2x^2y + xy^2 - x^2y + 2xy^2 - y^3$
    $= x^3 - 3x^2y + 3xy^2 - y^3$
19. $(x + y)(x^2 - xy + y^2)$
    $= x^3 - x^2y + xy^2 + x^2y - xy^2 + y^3$
    $= x^3 + y^3$
20. $(x - y)(x^2 + xy + y^2)$
    $= x^3 + x^2y + xy^2 - x^2y - xy^2 - y^3$
    $= x^3 - y^3$

Chapter 10 Simplifying Polynomial Expressions

Exercise 10

1. $8 + 2(x - 5)$
$= 8 + 2x - 10$
$= 2x - 2$

2. $-7(y - 4) + 9y$
$= -7y + 28 + 9y$
$= 2y + 28$

3. $10xy - x(5y - 3x) - 4x^2$
$= 10xy - 5xy + 3x^2 - 4x^2$
$= 5xy - x^2$

4. $(3x - 1)(2x - 5) + (x + 1)^2$
$= 6x^2 - 17x + 5 + x^2 + 2x + 1$
$= 7x^2 - 15x + 6$

5. $3x^2 - 4x - 5[x - 2(x - 8)]$
$= 3x^2 - 4x - 5[x - 2x + 16]$
$= 3x^2 - 4x - 5[-x + 16]$
$= 3x^2 - 4x + 5x - 80$
$= 3x^2 + x - 80$

6. $-x(x + 4) + 5(x - 2)$
$= -x^2 - 4x + 5x - 10$
$= -x^2 + x - 10$

7. $(a - 5)(a + 2) - (a - 6)(a - 4)$
$= a^2 - 3a - 10 - (a^2 - 10a + 24)$
$= a^2 - 3a - 10 - a^2 + 10a - 24$
$= 7a - 34$

8. $5x^2 - (-3xy - 2y^2)$
$= 5x^2 + 3xy + 2y^2$

9. $x^2 - [2x - x(3x - 1)] + 6x$
$= x^2 - [2x - 3x^2 + x] + 6x$
$= x^2 - [-3x^2 + 3x] + 6x$
$= x^2 + 3x^2 - 3x + 6x$
$= 4x^2 + 3x$

10. $(4x^2y^5)(-2xy^3)(-3xy) - 15x^2y^3(2x^2y^6 + 2)$
$= 24x^4y^9 - 30x^4y^9 - 30x^2y^3$
$= -6x^4y^9 - 30x^2y^3$

Chapter 11 Dividing Polynomials

Exercise 11

1. $\dfrac{15x^5 - 30x^2}{-5x}$

$= \dfrac{15x^5}{-5x} + \dfrac{-30x^2}{-5x}$

$= -3x^4 + 6x$

The quotient is $-3x^4 + 6x$ and the remainder is 0.

2. $\dfrac{-14x^4 + 21x^2}{-7x^2}$

$= \dfrac{-14x^4}{-7x^2} + \dfrac{21x^2}{-7x^2}$

$= 2x^2 - 3$

The quotient is $2x^2 - 3$ and the remainder is 0.

3. $\dfrac{25x^4y^2}{-5x} = -5x^3y^2$

The quotient is $-5x^3y^2$ and the remainder is 0.

4. $\dfrac{6x^5y^2 - 8x^3y^3 + 10xy^6}{2xy^2}$

$= \dfrac{6x^5y^2}{2xy^2} + \dfrac{-8x^3y^3}{2xy^2} + \dfrac{10xy^6}{2xy^2}$

$= 3x^4 - 4x^2y + 5y^4$

The quotient is $3x^4 - 4x^2y + 5y^4$ and the remainder is 0.

5. $\dfrac{-10x^4y^4z^4 - 20x^2y^5z^2}{10x^2y^3z}$

$= \dfrac{-10x^4y^4z^4}{10x^2y^3z} + \dfrac{-20x^2y^5z^2}{10x^2y^3z}$

$= -x^2yz^3 - 2y^2z$

The quotient is $-x^2yz^3 - 2y^2z$ and the remainder is 0.

6. $\dfrac{-18x^5 + 5}{3x^5}$

$= \dfrac{-18x^5}{3x^5} + \dfrac{5}{3x^5}$

$= -6 + \dfrac{5}{3x^5}$

The quotient is –6 and the remainder is 5.

7. $\dfrac{7a^6b^3 - 14a^5b^2 - 42a^4b^2 + 7a^3b^2}{7a^3b^2}$

$= \dfrac{7a^6b^3}{7a^3b^2} + \dfrac{-14a^5b^2}{7a^3b^2} + \dfrac{-42a^4b^2}{7a^3b^2} + \dfrac{7a^3b^2}{7a^3b^2}$

$= a^3b - 2a^2 - 6a + 1$

The quotient is $a^3b - 2a^2 - 6a + 1$ and the remainder is 0.

8. $\dfrac{x^2 - 1}{x + 1}$

$$
\begin{array}{r}
x - 1 \\
= x + 1\overline{\smash{\big)}\,x^2 + 0 - 1} \\
\underline{x^2 + x} \\
-x - 1 \\
\underline{-x - 1} \\
0
\end{array}
$$

The quotient is $x - 1$ and the remainder is 0.

9. $\dfrac{x^2 - 9x + 20}{x - 4}$

$$
\begin{array}{r}
x - 5 \\
= x - 4\overline{\smash{\big)}\,x^2 - 9x + 20} \\
\underline{x^2 - 4x} \\
-5x + 20 \\
\underline{-5x + 20} \\
0
\end{array}
$$

The quotient is $x - 5$ and the remainder is 0.

10. $\dfrac{2x^3 - 13x + x^2 + 6}{x - 4}$

$$
\begin{array}{r}
2x^2 + 9x + 23 \\
= x - 4\overline{\smash{\big)}\,2x^3 + x^2 - 13x + 6} \\
\underline{2x^3 - 8x^2} \\
9x^2 - 13x \\
\underline{9x^2 - 36x} \\
23x + 6 \\
\underline{23x - 92} \\
98
\end{array}
$$

The quotient is $2x^2 + 9x + 23$ and the remainder is 98.

Chapter 12 Factoring Polynomials

Exercise 12

1. False
2. False
3. False
4. False
5. False
6. $-a-b$
   $=-1a-1b$
   $=-1(a+b)$
   $=-(a+b)$
7. $-3x^2+6x-9$
   $=-3\cdot x^2-3\cdot -2x-3\cdot 3$
   $=-3(x^2-2x+3)$
8. $3-x$
   $=-1x+3$
   $=-1\cdot x -1\cdot -3$
   $=-1(x-3)$
   $=-(x-3)$
9. $24x^9y^2-6x^6y^7z^4$
   $=6x^6y^2\cdot 4x^3-6x^6y^2\cdot y^5z^4$
   $=6x^6y^2(4x^3-y^5z^4)$
10. $-45x^2+5$
    $=-5\cdot 9x^2-5\cdot -1$
    $=-5(9x^2-1)$
    $=-5(3x+1)(3x-1)$
11. $a^3b-ab+b$
    $=b(a^3-a+1)$
12. $14x+7y$
    $=7\cdot 2x+7\cdot y$
    $=7(2x+y)$
13. $x(2x-1)+3(2x-1)$
    $=(2x-1)(x+3)$
14. $y(a+b)+(a+b)$
    $=(a+b)(y+1)$
15. $x(x-3)+2(3-x)$
    $=x(x-3)-2(x-3)$
    $=(x-3)(x-2)$
16. $cx+cy+ax+ay$
    $=c(x+y)+a(x+y)$
    $=(x+y)(c+a)$
17. $x^2-3x-4=(x-4)(x+1)$
18. $x^2-49=(x+7)(x-7)$
19. $6x^2+x-15=(3x+5)(2x-3)$
20. $16x^2-25y^2=(4x+5y)(4x-5y)$
21. $27x^3-64=(3x-4)(9x^2+12x+16)$
22. $8a^3+125b^3=(2a+5b)(4a^2-10ab+25b^2)$
23. $2x^4y^2z^3-32x^2y^2z^3$
    $=2x^2y^2z^3\cdot x^2-2x^2y^2z^3\cdot 16$
    $=2x^2y^2z^3(x^2-16)$
    $=2x^2y^2z^3(x+4)(x-4)$
24. $a^2(a+b)-2ab(a+b)+b^2(a+b)$
    $=(a+b)(a^2-2ab+b^2)$
    $=(a+b)(a-b)^2$

Chapter 13 Rational Expressions

Exercise 13

1. $\dfrac{18x^3y^4z^2}{54x^3z^2}$

$= \dfrac{18x^3z^2 \cdot y^4}{18x^3z^2 \cdot 3}$

$= \dfrac{\cancel{18x^3z^2} \cdot y^4}{\cancel{18x^3z^2} \cdot 3}$

$= \dfrac{y^4}{3}$

2. $\dfrac{15y}{3y}$

$= \dfrac{3y \cdot 5}{3y \cdot 1}$

$= \dfrac{5}{1}$

$= 5$

3. $\dfrac{x-5}{5-x}$

$= \dfrac{1(x-5)}{-1(x-5)}$

$= \dfrac{1\cancel{(x-5)}}{-1\cancel{(x-5)}}$

$= \dfrac{1}{-1}$

$= -1$

4. $\dfrac{4a}{4+a}$ is simplified.

5. $\dfrac{2x-6}{x^2-5x+6}$

$= \dfrac{2(x-3)}{(x-2)(x-3)}$

$= \dfrac{2\cancel{(x-3)}}{(x-2)\cancel{(x-3)}}$

$= \dfrac{2}{x-2}$

6. $\dfrac{x^2-4}{x^2+4x+4}$

$= \dfrac{(x+2)(x-2)}{(x+2)(x+2)}$

$= \dfrac{\cancel{(x+2)}(x-2)}{\cancel{(x+2)}(x+2)}$

$= \dfrac{x-2}{x+2}$

7. $\dfrac{x(a+b)+y(a+b)}{x+y}$

$= \dfrac{(a+b)(x+y)}{(x+y)}$

$= \dfrac{(a+b)\cancel{(x+y)}}{\cancel{(x+y)}}$

$= \dfrac{a+b}{1}$

$= a + b$

8. $\dfrac{7x}{35x-14}$

$= \dfrac{7 \cdot x}{7(5x-2)}$

$= \dfrac{\cancel{7} \cdot x}{\cancel{7}(5x-2)}$

$= \dfrac{x}{5x-2}$

9. $\dfrac{4x^2y - 4xy - 24y}{2x^2 - 18}$

$= \dfrac{4y\left(x^2 - x - 6\right)}{2\left(x^2 - 9\right)}$

$= \dfrac{2 \cdot 2y(x+2)(x-3)}{2(x+3)(x-3)}$

$= \dfrac{\cancel{2(x-3)} \cdot 2y(x+2)}{\cancel{2(x-3)}(x+3)}$

$= \dfrac{2y(x+2)}{x+3}$

10. $\dfrac{x - y}{x^3 - y^3}$

$= \dfrac{(x-y) \cdot 1}{(x-y)\left(x^2 + xy + y^2\right)}$

$= \dfrac{1}{x^2 + xy + y^2}$

11. $\dfrac{x^2 - 4x + 4}{x^2 - 9} \cdot \dfrac{2x - 6}{x - 2}$

$= \dfrac{(x-2)(x-2)}{(x+3)(x-3)} \cdot \dfrac{2(x-3)}{(x-2)}$

$= \dfrac{\cancel{(x-2)}(x-2)}{(x+3)\cancel{(x-3)}} \cdot \dfrac{2\cancel{(x-3)}}{\cancel{(x-2)}}$

$= \dfrac{2(x-2)}{x+3}$

12. $\dfrac{x-1}{2x-1} \div \dfrac{x+1}{4x-2}$

$= \dfrac{x-1}{2x-1} \cdot \dfrac{4x-2}{x+1}$

$= \dfrac{(x-1)}{(2x-1)} \cdot \dfrac{2(2x-1)}{(x+1)}$

$= \dfrac{(x-1)}{\cancel{(2x-1)}} \cdot \dfrac{2\cancel{(2x-1)}}{(x+1)}$

$= \dfrac{2(x-1)}{x+1}$

13. $\dfrac{2}{x^2 - 2x - 3} + \dfrac{4}{x-3}$

$= \dfrac{2}{(x+1)(x-3)} + \dfrac{4}{(x-3)}$

$= \dfrac{2}{(x+1)(x-3)} + \dfrac{4(x+1)}{(x+1)(x-3)}$

$= \dfrac{2}{(x+1)(x-3)} + \dfrac{4x+4}{(x+1)(x-3)}$

$= \dfrac{4x+6}{(x+1)(x-3)}$

$= \dfrac{2(2x+3)}{(x+1)(x-3)}$

14. $\frac{2x}{x^2-14x+49}-\frac{1}{x-7}$

$=\frac{2x}{(x-7)(x-7)}-\frac{1}{(x-7)}$

$=\frac{2x}{(x-7)(x-7)}-\frac{1(x-7)}{(x-7)(x-7)}$

$=\frac{2x}{(x-7)(x-7)}-\frac{x-7}{(x-7)(x-7)}$

$=\frac{2x-(x-7)}{(x-7)(x-7)}$

$=\frac{2x-x+7}{(x-7)(x-7)}$

$=\frac{x+7}{(x-7)(x-7)}$

15. $\frac{\frac{2}{3x}}{\frac{1}{x+1}}$

$=\frac{2}{3x}\div\frac{1}{x+1}$

$=\frac{2}{3x}\cdot\frac{x+1}{1}$

$=\frac{2x+2}{3x}$

Chapter 14 Solving Linear Equations and Inequalities

Exercise 14

1. $x-7=11$

$x-7+7=11+7$

$x=18$

2. $6x-3=13$

$6x-3+3=13+3$

$6x=16$

$x=\frac{16}{6}=\frac{8}{3}$

3. $x+3(x-2)=2x-4$

$x+3x-6=2x-4$

$4x-6=2x-4$

$4x-6-2x=2x-4-2x$

$2x-6=-4$

$2x-6+6=-4+6$

$2x=2$

$x=1$

4. $\frac{x+3}{5}=\frac{x-1}{2}$

$\frac{10}{1}\cdot\frac{x+3}{5}=\frac{10}{1}\cdot\frac{x-1}{2}$

$2(x+3)=5(x-1)$

$2x+6=5x-5$

$2x+6-5x=5x-5-5x$

$-3x+6=-5$

$-3x+6-6=-5-6$

$-3x=-11$

$\frac{-3x}{-3}=\frac{-11}{-3}$

$x=\frac{11}{3}$

5.
$$\begin{aligned} 3x + 2 &= 6x - 4 \\ 3x + 2 - 6x &= 6x - 4 - 6x \\ -3x + 2 &= -4 \\ -3x + 2 - 2 &= -4 - 2 \\ -3x &= -6 \\ x &= 2 \end{aligned}$$

6. Solve for y: $-12x + 6y = 9$
$$\begin{aligned} -12x + 6y &= 9 \\ -12x + 6y + 12x &= 9 + 12x \\ 6y &= 9 + 12x \\ \frac{6y}{6} &= \frac{9+12x}{6} \\ y &= \frac{3}{2} + 2x \end{aligned}$$

7.
$$\begin{aligned} -x + 9 &< 0 \\ -x + 9 - 9 &< 0 - 9 \\ -x &< -9 \\ \frac{-x}{-1} &> \frac{-9}{-1} \\ x &> 9 \end{aligned}$$

8.
$$\begin{aligned} 3x + 2 &> 6x - 4 \\ 3x + 2 - 6x &> 6x - 4 - 6x \\ -3x + 2 &> -4 \\ -3x + 2 - 2 &> -4 - 2 \\ -3x &> -6 \\ \frac{-3x}{-3} &< \frac{-6}{-3} \\ x &< 2 \end{aligned}$$

9.
$$\begin{aligned} 3x - 2 &\le 7 - 2x \\ 3x - 2 + 2x &\le 7 - 2x + 2x \\ 5x - 2 &\le 7 \\ 5x - 2 + 2 &\le 7 + 2 \\ 5x &\le 9 \\ x &\le 1.8 \end{aligned}$$

10.
$$\begin{aligned} \frac{x+3}{5} &\ge \frac{x-1}{2} \\ \frac{10}{1}\cdot\frac{x+3}{5} &\ge \frac{10}{1}\cdot\frac{x-1}{2} \\ 2(x + 3) &\ge 5(x - 1) \\ 2x + 6 &\ge 5x - 5 \\ 2x + 6 - 5x &\ge 5x - 5 - 5x \\ -3x + 6 &\ge -5 \\ -3x + 6 - 6 &\ge -5 - 6 \\ -3x &\ge -11 \\ \frac{-3x}{-3} &\le \frac{-11}{-3} \\ x &\le \frac{11}{3} \end{aligned}$$

Chapter 15 Solving Quadratic Equations

Exercise 15

1. $x^2 - x - 6 = 0$
   $(x - 3)(x + 2) = 0$
   $x - 3 = 0$ or $x + 2 = 0$
   $x = 3$ or $x = -2$

2. $x^2 + 6x + 9 = -5 + 9$
   $(x + 3)^2 = 4$
   $x + 3 = \pm 2$
   $x + 3 = 2$ or $x + 3 = -2$
   $x = -1$ or $x = -5$

3. $3x^2 - 5x + 1 = 0$
   $$x = \frac{-(-5) \pm \sqrt{(-5)^2 - 4(3)(1)}}{2(3)}$$
   $$x = \frac{5 \pm \sqrt{13}}{6}$$

4. $x^2 - 6 = 8$
   $x^2 = 14$
   $x = \pm\sqrt{14}$

5. $x^2 - 3x + 2 = 0$
   $$x = \frac{-(-3) \pm \sqrt{(-3)^2 - 4(1)(2)}}{2(1)}$$
   $$= \frac{3 \pm \sqrt{9 - 8}}{2}$$
   $$x = \frac{3 \pm \sqrt{1}}{2}$$
   $$x = \frac{3 \pm 1}{2}$$
   $$x = \frac{4}{2} = 2 \text{ or } x = \frac{2}{2} = 1$$

6. $9x^2 + 18x - 17 = 0$
   $$x = \frac{-18 \pm \sqrt{18^2 - 4(9)(-17)}}{2(9)}$$
   $$x = \frac{-18 \pm \sqrt{936}}{18}$$
   $$x = \frac{-18 \pm \sqrt{36(26)}}{18}$$
   $$x = \frac{-18 \pm 6\sqrt{26}}{18}$$
   $$x = \frac{-3 \pm \sqrt{26}}{3}$$

7. $6x^2 - 12x + 7 = 0$
   $$x = \frac{-(-12) \pm \sqrt{(-12)^2 - 4(6)(7)}}{2(6)}$$
   $$= \frac{12 \pm \sqrt{144 - 168}}{12}$$
   $$x = \frac{12 \pm \sqrt{-24}}{12}$$
   There is no real solution because the discriminant is negative.

8. $x^2 - 10x = -25$
   $x^2 - 10x + 25 = -25 + 25$
   $(x - 5)^2 = 0$
   $x = 5$

9. $-x^2 = -9$
   $x^2 = 9$
   $x = \pm 3$

10. $6x^2 = x + 2$
    $6x^2 - x - 2 = 0$
    $(2x + 1)(3x - 2) = 0$
    $2x + 1 = 0$ or $3x - 2 = 0$
    $x = -\frac{1}{2}$ or $x = \frac{2}{3}$

Chapter 16 The Cartesian Coordinate Plane

Exercise 16

1. True
2. False
3. True
4. False
5. True
6. False
7. rise
8. run
9. negative
10. positive
11. zero
12. $\frac{2}{3}$
13. $-\frac{4}{3}$
14. undefined

15.

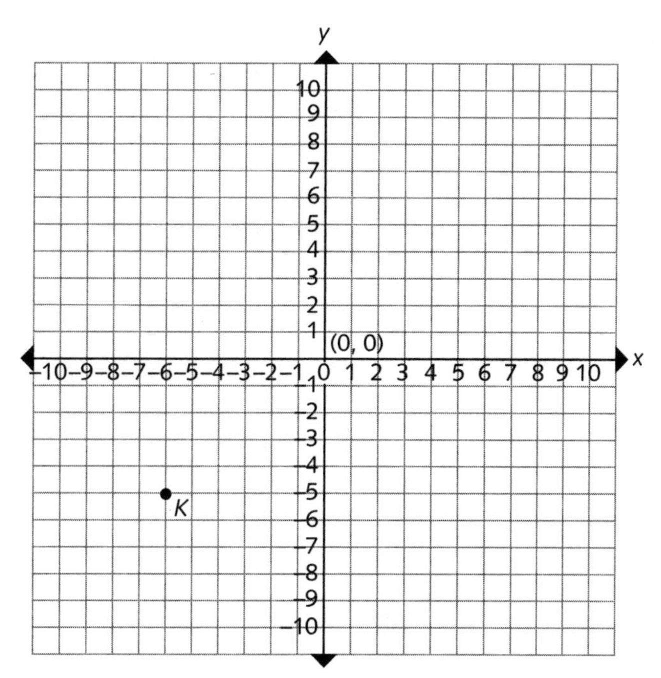

The point K is 6 units to the left of the y-axis and 5 units below the x-axis, so (–6, –5) is the ordered pair corresponding to point K.

16. $d = \sqrt{(5-1)^2 + (7-4)^2}$

$d = \sqrt{16+9} = \sqrt{25} = 5$

17. $d = \sqrt{(4+2)^2 + (-1-5)^2}$

$d = \sqrt{36+36} = \sqrt{2(36)} = 6\sqrt{2}$

18. Midpoint $= \left(\dfrac{-2+4}{2}, \dfrac{5-1}{2}\right) = (1, 2)$

19. $m = \dfrac{5+1}{-2-4} = \dfrac{6}{-6} = -1$

20.

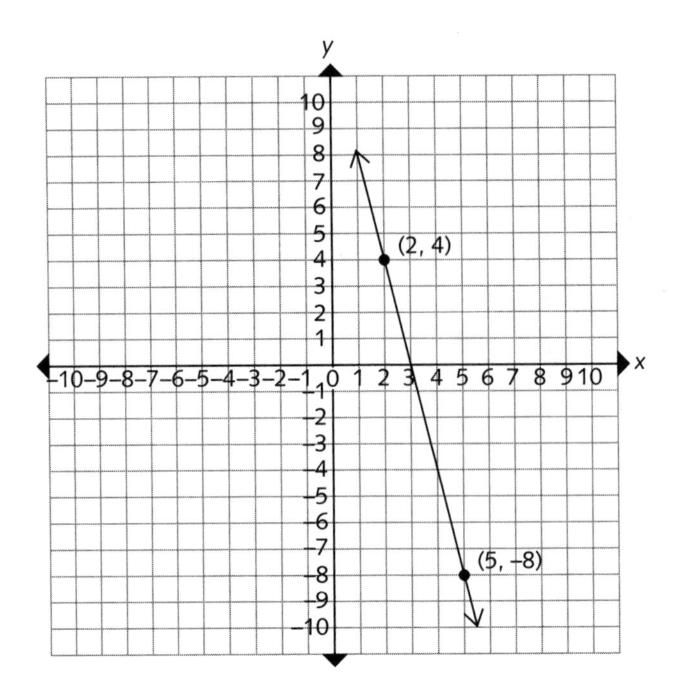

$$m = \frac{4+8}{2-5} = \frac{12}{-3} = -4$$

Chapter 17 Graphing Linear Equations

Exercise 17

1. $m = \frac{14-11}{3-5} = \frac{3}{-2} = -\frac{3}{2}$
2. $y = -2x + 6$

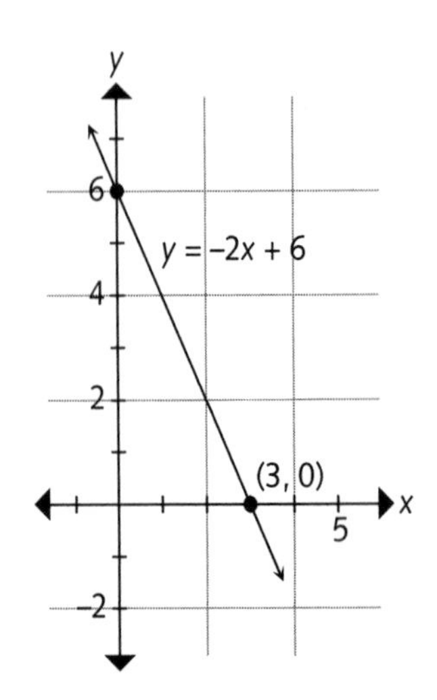

3. $3y = 5x - 9$

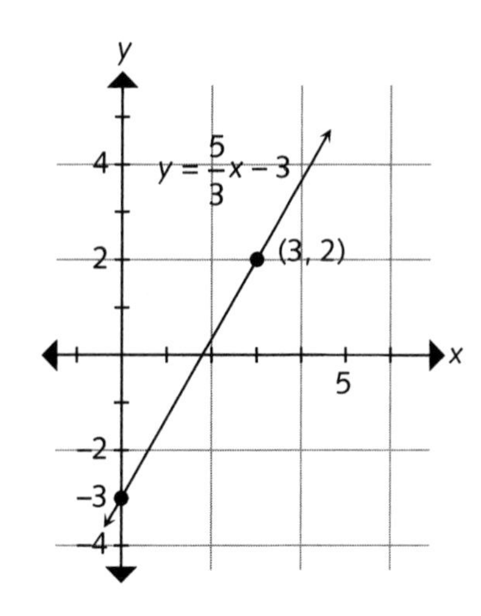

4. $y = x$

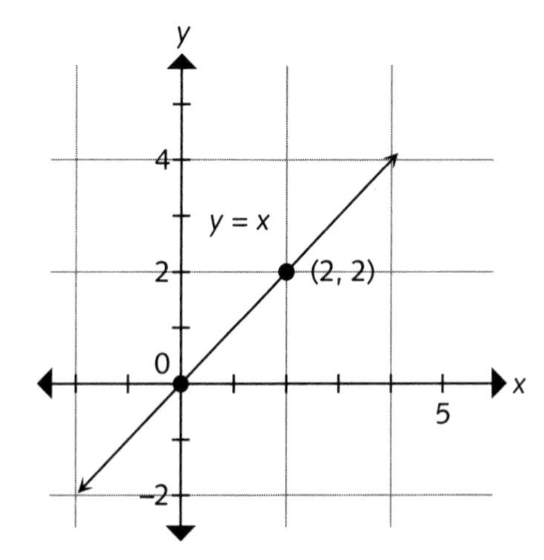

5. $4y - 5x = 8$

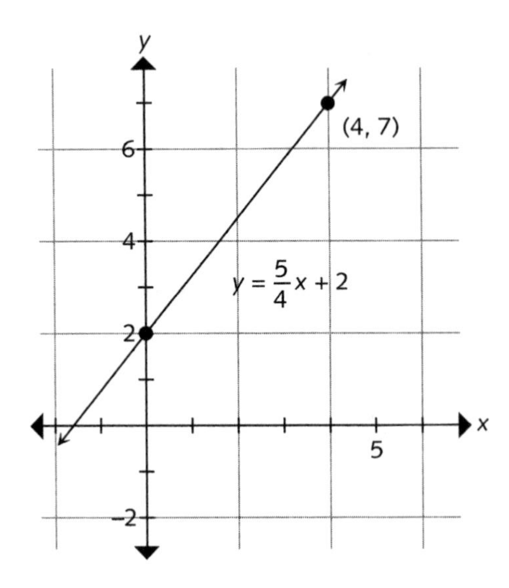

6. $y = \frac{1}{3}x - \frac{2}{3}$

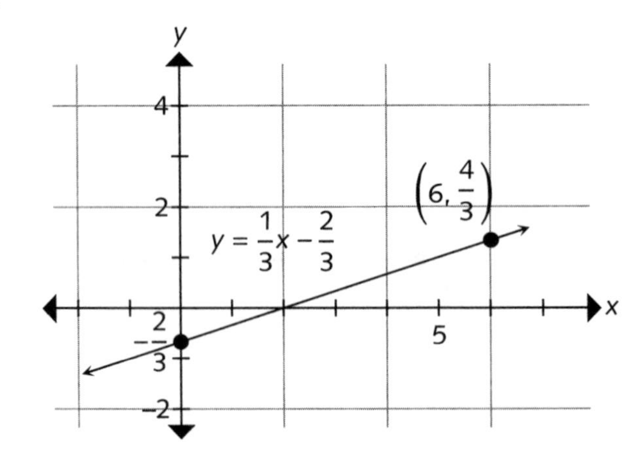

Chapter 18 The Equation of a Line

Exercise 18

1. $y = 4x + 3$
2. $y = -3x - 3$
3. $y = \frac{1}{3}x$
4. $$2 = \frac{y-1}{x-1}$$ $$2(x - 1) = y - 1$$ $$y = 2x - 1$$

5. $$-1 = \frac{y-3}{x-2}$$
$-1(x - 2) = y - 3$
$-x + 2 = y - 3$
$y = -x + 5$

6. $$\frac{1}{5} = \frac{y-1}{x}$$
$x = 5y - 5$
$5y = x + 5$
$$y = \frac{1}{5}x + 1$$

7. $$\frac{y-4}{x-2} = \frac{4-2}{2-1}$$
$$\frac{y-4}{x-2} = 2$$
$y - 4 = 2(x - 2)$
$y - 4 = 2x - 4$
$y = 2x$

8. $$\frac{y-2}{x-1} = \frac{2-2}{-1-1} = 0$$
$y - 2 = 0$
$y = 2$

9. $$\frac{y-1}{x} = \frac{-1}{2-1} = -1$$
$y - 1 = -x$
$y = -x + 1$

10. $$\frac{6-4}{6-4} = \frac{y-4}{x-4}$$
$$1 = \frac{y-4}{x-4}$$
$y - 4 = x - 4$
$y = x$

Chapter 19 Basic Function Concepts

Exercise 19

1. a. $f = \{(2, 1), (4, 5), (6, 9), (5, 9)\}$
   b. $g = \{(3, 4), (5, 1), (6, 3), (3, 6)\}$
   c. $h = \{(2, 1)\}$
   d. $t = \{(7, 5), (8, 9), (8, 9)\}$
   Only f, h, and t are functions. Note that in t, (8, 9) and (8, 9) are the same point.
2. The domain is {4, 6, 7, 8} and the range is {5, 7, 9}.
3. a. $y = f(x) = 5x - 7$ The domain is the set of all real numbers.
   b. $y = g(x) = \sqrt{2x-3} + 4$
   Set $2x - 3 \geq 0$ and solve.
   $2x - 3 \geq 0$
   $2x \geq 3$
   $x \geq \frac{3}{2}$. The domain is the set of all real numbers greater than or equal to $\frac{3}{2}$.

c. $y = \frac{9x+1}{x-5}$ The domain is the set of all real numbers except 5.

d. $y = \frac{2x^2+5}{x^2-4}$

Set $x^2 - 4 = 0$ and solve.

$x = \pm 2$ The domain is the set of all real numbers except 2 and −2.

4. a. $f(2) = 5\sqrt{2+2} - 3 = 5\sqrt{4} - 3 = 10 - 3 = 7$
   b. $f(-1) = 5\sqrt{-1+2} - 3 = 5\sqrt{1} - 3 = 5 - 3 = 2$
   c. $f(6) = 5\sqrt{6+2} - 3 = 5\sqrt{8} - 3 = 5\sqrt{4(2)} - 3 = 10\sqrt{2} - 3$
   d. $f(-3) = 5\sqrt{-3+2} - 3 = 5\sqrt{-1} - 3$ There is no real number solution because the square root of a negative number is not a real number.
5. Only graphs b and c are functions.
6. $y = 4x + 1$

Chapter 20 Systems of Equations

Exercise 20

1. $x - 2y = -4$
   $2x + y = 7$
   $x = 2y - 4$
   $2(2y - 4) + y = 7$
   $4y - 8 + y = 7$
   $5y = 15$
   $y = 3$
   $2x + 3 = 7$
   $2x = 4$
   $x = 2$
   $x = 2$ and $y = 3$ is the solution.

2. $4x - y = 3$
   $x - 3y = -13$
   $y = 4x - 3$
   $x - 3(4x - 3) = -13$
   $x - 12x + 9 = -13$
   $-11x + 9 = -13$
   $-11x = -22$
   $x = 2$
   $4(2) - y = 3$
   $8 - y = 3$
   $-y = -5$
   $y = 5$
   $x = 2$ and $y = 5$ is the solution.

3. $4x + 2y = 8$

$2x - 3y = -8$

$2y = -4x + 8$

$y = -2x + 4$

$2x - 3(-2x + 4) = -8$

$2x + 6x - 12 = -8$

$8x = 4$

$x = \frac{1}{2}$

$4\left(\frac{1}{2}\right) + 2y = 8$

$2 + 2y = 8$

$2y = 6$

$y = 3$

$x = \frac{1}{2}$ and $y = 3$ is the solution.

4. $-2x + 4y = 8 \longrightarrow -2x + 4y = 8$

$-2x - y = -7 \xrightarrow[\text{by } -1]{\text{Multiply}} 2x + y = 7$

$5y = 15$

$y = 3$

$-2x - 3 = -7$

$-2x = -4$

$x = 2$

$x = 2$ and $y = 3$ is the solution.

5. $8x - 2y = 6 \longrightarrow 8x - 2y = 6$

$x - 3y = -13 \xrightarrow[\text{by } -8]{\text{Multiply}} -8x + 24y = 104$

$22y = 110$

$y = 5$

$x - 3(5) = -13$

$x - 15 = -13$

$x = 2$

$x = 2$ and $y = 5$ is the solution.

6. $2x + y = 4 \xrightarrow[\text{by } -2]{\text{Multiply}} -4x - 2y = -8$

$4x - 6y = -16 \longrightarrow 4x - 6y = -16$

$-8y = -24$

$y = 3$

$2x + 3 = 4$

$2x = 1$

$x = \frac{1}{2}$

$x = \frac{1}{2}$ and $y = 3$ is the solution.

7. 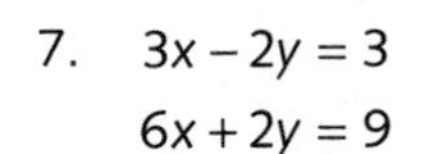

$3x - 2y = 3$

$6x + 2y = 9$

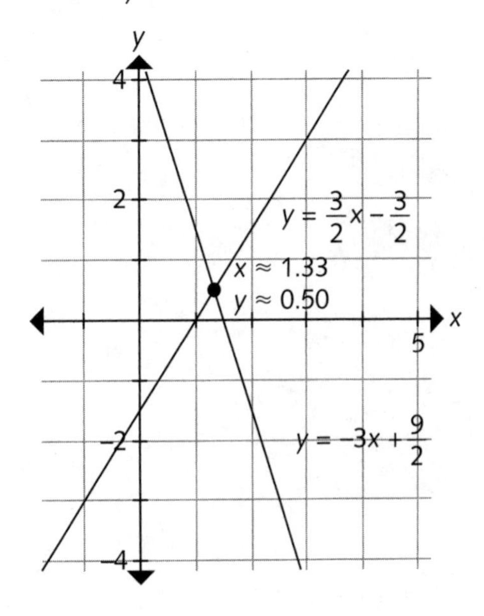

8. $7x + 14y = 2$
   $14x - 7y = -11$

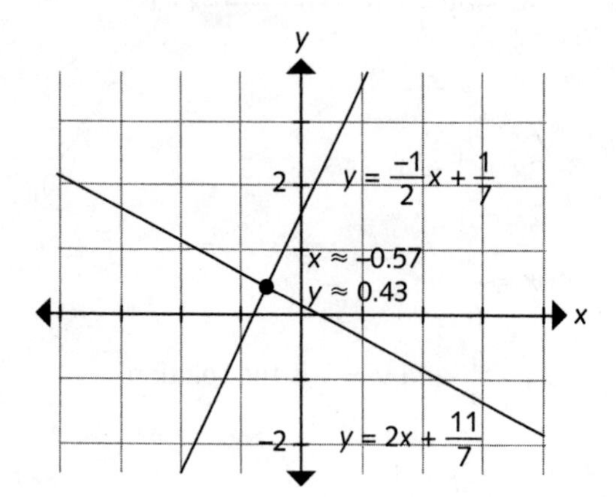

Index

Page numbers followed by *f* indicate material in figures. Page numbers followed by *t* indicate material in tables.

Absolute value
  of coordinates, 171–173
  definition of, 15
  of negative numbers, 15–17, 15f
  on number line, 15, 15f
  in order of operations
    algebraic expressions, 68
    with multiplication, 120
    PEMDAS, 60, 62
    with subtraction, 120
  of real numbers, 15–17
  square root and, 32, 36, 164
Absolute value bars, 15, 36, 58, 60
Absolute value function, 202, 202f
Addition
  of algebraic fractions, 146–150
  associative property of, 9, 10
  closure property of, 8
  commutative property of, 9
  of decimals, 20
  distributive property and, 11
  division and, 59, 120
  in factoring, 134–136
  of fractions, 19–20
  within fractions
    complex, 151–152
    order of operations, 59
    in rational expressions, 142–150
    simplifying and, 120
  of integers, 4
  of like terms, 88–89
  linear equation rules, 155
  linear inequality rules, 161
  of monomials, 88–89
  of natural numbers, 1
  of negative numbers, 17–20
  of opposite number, 10
  in order of operations
    algebraic expressions, 68–69
    with exponentiation, 48, 120
    PEMDAS, 58–63
    with square root symbol, 35, 119
  of polynomials, 90–91
  power of a sum rule, 80, 81, 119–120, 220
  sign for, 8
  signed numbers rules, 17–20
  in special products, 102
  of whole numbers, 2
  of zero, 10, 18
Additive identity property, 10, 11
Additive inverse property, 10–11
Algebraic expressions
  definition of, 66
  evaluating, 66–70

Algebraic expressions (*cont.*)
  as GCF, 121–126
  parentheses in, 67–72
  terms in. *See* Terms
Algebraic fractions
  adding, 146–150
  dividing, 145–146
  multiplying, 143–145
  reducing to lowest terms, 139–143
  subtracting, 146–150
"Approximately equal to" symbol (≈), 5
Associative property of addition, 9, 10
Associative property of multiplication, 10

Base
  definition of, 44
  in exponential expression, 45f
  product rule and, 74–75
  quotient rule and, 75–77
Binary operation, 8
Binomials
  definition of, 85
  in factoring, 134–136
  identifying, 86
  multiplying, 98–101, 127–131
  special products, 102
Braces, 58
Brackets, 58, 108

Cartesian coordinate plane
  functions graphed in, 201
  number lines in, 171, 172f
  ordered pairs in. *See* Ordered pairs
  origin of, 171, 172f, 191
  quadrants of, 174–176, 175f
Circles, 6, 64–65, 203, 217–218
Closure property of addition, 8
Closure property of multiplication, 8–9
Coefficients
  GCF with negative, 123–125
  linear equation rules, 154
  linear inequality rules, 161
  monomials and, 84–85, 87–88, 94–98
  one as, 65
  in quadratic trinomials, 132–133, 168
  in radical simplifying, 40–42
  of variables, 65–66
Commutative property of addition, 9
Commutative property of multiplication, 9
Commutative property of subtraction, 24
Completing the square technique, 166–167
Complex fractions, 151–152
Complex numbers, 163, 167
Constants
  definition of, 64
  determining, 64–65
  grouping symbols and, 66
  in linear equations, 154
  in monomials, 84–85, 87
Coordinate axes, 171, 172f
Coordinate of a number line point, 2
Coordinates
  absolute value of, 171–173
  determining, 173, 173f
  diagram of, 172f
  in function set, 195–196
  order of, 171
  at origin, 171
  in quadrants, 174–175, 175f
  x-coordinate. *See* x-coordinate
  y-coordinate. *See* y-coordinate
Counting numbers. *See* Natural numbers
Cube root
  categorization of, 5–7
  of decimals, 37, 39
  definition of, 36
  function with, 200
  in order of operations, 37
  principal, 36–37, 39, 53

Decimals
  absolute value of, 16
  addition of, 20
  categorization of, 4, 6–7
  cube root of, 37, 39
  exponentiation of
    with fractions, 53
    with natural numbers, 47
    with negative numbers, 51
    with one, 45
    with zero, 49
  on number line, 6
  perfect square, 33
  repeating, 5
  rounding of, 5
  square of, 47
  square root of, 34, 53
  terminating, 4–5

Dependent variable, 197
Difference of two cubes, 102, 134, 136
Difference of two squares, 102, 134–135
Distance between two points in a plane, 176–177
Distributive property, 11
Division. *See also* Fractions
addition and, 59, 120
of algebraic fractions, 145–146
of complex fractions, 151
components of, 110
with exponents
fractions, 54–56
natural numbers, 47–48
negative numbers, 51–52, 120
one, 45
power of a quotient rule, 79
quotient rule for, 75–77, 81
zero, 49
by GCF, 139–143
of integers, 4
linear equation rules, 155
linear inequality rules, 159, 161
multiplication and, 120
of natural numbers, 1–2
negative numbers in, 29–30
in order of operations, 59–63, 68
of polynomials, 110–117, 120
in polynomial expressions, 104–105
quotient rule for, 75–77, 81
sign for, 29
signed numbers rules, 29–30
simplifying, 120
of whole numbers, 2
zero in, 4, 29, 110, 139, 155, 180–181
Domain of functions, 195–200

e (transcendental number), 6, 8
Elimination method, 208–210
"Equal to" symbol (=), 120
Equations
linear. *See* Linear equations
quadratic. *See* Quadratic equations
sides of, 154
solving, 154
Exponents
definition of, 44
in exponential expression, 45f
fractions as, 53–57, 84–85
highest common, 121
natural numbers as, 44–48, 74
negative numbers as, 50–52, 77,
104–105, 120
one as, 45, 95
rational, 54–57
rules for, 74–81
zero as, 48–49
Exponential expression
components of, 45f
definition of, 44
parentheses in, 48, 59
power of a product rule, 78–80, 119–120
power of a quotient rule, 79
a power to a power rule, 77–78
product rule, 74–75
quotient rule, 75–77
reciprocals of, 50–51
Exponentiation
definition of, 44
in order of operations
with addition, 48, 120
algebraic expressions, 69
PEMDAS, 59–62
polynomial expressions, 106–108

Factors
"equal to" symbol and, 120
greatest common, 121–126, 139–143
prime, 148
vs. terms, 119
Factoring
of algebraic fractions, 139–143
definition of, 119
by FOIL method, 127–131
by grouping, 126–127, 131–133
guidelines for, 137
with negative coefficients, 123–125
objective of, 119
one in, 122–126
perfect trinomial squares, 133–134
quadratic trinomials, 127–133
two terms, 134–136
Fahrenheit to Celsius conversion, 70
Fifth root, 38
FOIL method, 99–101, 127–131
Fourth root, 6, 38
Fractions
absolute value of, 16–17
addition of, 19–20

ons

, 120

quotient rule, 79
quotient rule for, 75–77, 81
with zero, 49
in linear equations, 158
as monomials, 84–85
on number line, 6
in order of operations, 59
order of operations in, 59
perfect square, 33
in radical simplifying, 39
in rational expressions, 140–150
signed numbers rules, 29–30
simplifying, 120
square root of, 34–35, 41–42
Fraction bars, 29, 58, 59, 151
Functional relationships, 203
Functions, 195–202, 202f

GCF, 121–126, 139–143
General polynomial, 85, 86
Graph of a number, 2, 2f
Graphing method, 210–211
"Greater than" symbol (>), 14t, 159
"Greater than or equal to" symbol (>), 14t, 159
Greatest common factor (GCF), 121–126, 139–143
Grouping symbols
absolute value bars, 15, 36, 58, 60
braces, 58
brackets, 58, 108
constants and, 66
fraction bars, 29, 58, 59, 151
in order of operations
with addition, 35, 119
algebraic expressions, 67–70
with multiplication, 36, 37, 39–42
PEMDAS, 58–62
polynomial expressions, 106–108
parentheses. *See* Parentheses
purpose of, 58
square root. *See* Square root symbol
variables and, 66

Horizontal axis, 171, 172f

Independent variables, 197, 198
Index, 38
Inequality symbols, 14t, 120
Inputs (domain value), 197
Integers, 3–5, 3f, 6f
Irrational numbers, 5–6, 6f

Least common denominator (LCD), 148–149, 152
Least common multiple, 158
"Less than" symbol (<), 14t, 159
"Less than or equal to" symbol (<), 14t, 159
Like terms, 87, 88–90
Linear equations
constants in, 154
fractions in, 158
graphing, 184–188
one-variable, 154–158
point-slope form of, 190–193
sides of, 154
slope *y*-intercept form of, 184–190
solving two simultaneous, 205–211
tools for solving, 155
two-variable, 159
Linear functions, 201, 202f
Linear inequalities, 159–161
Lines
parallel, 181
perpendicular, 181–182
point-slope equation for, 190–193
properties of, 184
slope of. *See* Slope of a line
slope *y*-intercept equation for, 184–190
standard form of equation for, 184

Middle terms, 99–101, 127–128, 131, 133
Midpoint between two points in a plane, 177
Minus sign (–)
in factoring, 124
parentheses and, 71
rules for, 25
in subtraction, 21, 22, 92
terms and, 83–84

Monomials
  addition of, 88–89
  definition of, 84
  division of polynomials by, 110–113
  GCF, 121–126, 139–143
  identifying, 84–86
  like terms, 87, 88–90
  multiplying, 94–98
  subtraction of, 88–89
  unlike terms, 87
Multiplication
  in algebraic expressions, 66
  of algebraic fractions, 143–145
  associative property of, 10
  closure property of, 8–9
  commutative property of, 9
  distributive property and, 11
  division and, 120
  factoring and, 119
  within fractions, 120, 140, 143
  of integers, 4
  linear equation rules, 155
  linear inequality rules, 159–160
  of monomials, 94–98
  of natural numbers, 1
  by one, 10
  in order of operations
    with absolute value, 120
    algebraic expressions, 67–70
    PEMDAS, 60–62
    polynomial expressions, 106–109
    with square root symbol, 36, 37, 39–42
  parentheses in, 8, 58, 144
  of polynomials, 94–102
  power of a product rule for, 78–80, 119–120
  power to a power rule for, 77–78
  product rule for, 74–75
  by reciprocal, 11
  signed numbers rules for, 26–28
  symbol for, 8
  of whole numbers, 2
  by zero, 11, 12, 26, 28, 155
Multiplicative identity property, 10, 11
Multiplicative inverse property, 11

Natural numbers
  addition of, 1
  definition of, 1
  division of, 1–2
  as exponent, 44–48, 74–81
  multiplication of, 1
  on number line, 1, 1f
  rational. *See* Rational numbers
  roots of, 38
  set of, 1
  subtraction of, 1
  zero and, 2
Negative integer, 3, 3f
Negative numbers
  absolute value of, 15–17, 15f
  addition of, 17–20
  categorization of, 7
  comparing, 15
  cube of, 36
  cube root of, 5–6, 36–37, 39, 53
  in division, 29–30
  domain of functions and, 196–197
  even roots of, 6, 39, 53–54, 196–197
  as exponent, 50–52, 77, 84–85, 104–105, 120
  exponentiation of
    with fractions, 53–56
    with natural numbers, 46–47
    with negative numbers, 50, 52
    with one, 45
    with zero, 49
  fifth root of, 38
  as GCF, 123–125
  integers, 3, 3f
  linear inequality rules, 159–160
  in monomials, 84
  multiplication of, 26–28
  *n*th root of, 38
  on number line, 3–4, 3f, 6, 15, 160f
  parentheses with, 25, 58, 67, 176, 178
  seventh root of, 39
  square of, 32
  square root of, 6, 32, 36, 163
  subtraction of, 21–25
Negative sign (–)
  in division of polynomials, 111
  in factoring, 132
  in quadratic formula, 168
  rules for, 25
  when combining like terms, 90
Nonintegers, 6
"Not equal to" symbol (≠), 14t, 120
*n*th root of *a*, 38

Numbers
- complex, 163, 167
- integers, 3–5, 3f, 6f
- irrational, 5–6, 6f
- natural. *See* Natural numbers
- opposite. *See* Opposite numbers
- perfect cubes, 36, 102
- perfect squares. *See* Perfect squares
- rational. *See* Rational numbers
- real. *See* Real numbers
- as variables, 64
- whole. *See* Whole numbers

Number lines
- absolute value on, 15, 15f
- in Cartesian coordinate plane, 171, 172f
- decimals on, 6
- *e* on, 8
- fractions on, 6
- graph of a number on, 2, 14
- integers on, 3–4, 3f
- natural numbers on, 1, 1f
- negative numbers on, 3–4, 3f, 6, 15, 160f
- opposite numbers on, 2–4, 3f
- pi on, 6
- positive numbers on, 3–4, 3f, 6, 160f
- real numbers on, 6, 6f, 8
- square root on, 6
- whole numbers on, 2, 2f, 3–4, 3f

Numerical coefficient. *See* Coefficients

One
- categorization of
  - as integer, 3
  - as multiplicative identity, 10
  - as natural number, 1
  - as perfect cube, 36
  - as perfect square, 33
  - as rational number, 4
  - as whole number, 2
- as coefficient, 65
- cube root of, 36–37
- as exponent, 45, 95
- exponentiation of, 48, 49
- in factoring, 122–126
- as GCF, 142
- multiplication by, 10
- in slope y-intercept equation, 185–186
- square root of, 33
- in x-y T-table, 185

One-variable linear equations, 154–158

Opposite numbers
- as additive inverse, 10–11
- categorization of, 3
- in elimination method, 208–210
- in factoring, 123–125
- on number line, 2–4, 3f
- in subtraction, 21–25, 92–93

Opposite symbol (–), 25, 71

Ordered pairs
- determining, 173, 173f
- equal, 173–174
- function set of, 195–196, 201
- points. *See* Points
- written order of, 171
- zero in, 174

Origin, 171, 172f, 191

Outputs (range value), 197

Parallel lines, 181

Parentheses
- minus sign and, 71
- in multiplication, 8, 58, 144
- with negative numbers, 25, 58, 67, 176, 178
- opposite symbol and, 71
- in order of operations
  - in algebraic expressions, 67–72
  - within brackets, 108
  - in exponential expression, 48, 59
  - in PEMDAS, 58–62
  - in polynomial expressions, 106–108
- in subtraction, 147

Partial products, 99–101

PEMDAS, 60–63

Percent, 4

Perfect cubes, 36, 102

Perfect squares
- definition of, 33
- multiplying, 102
- principal square roots list, 33
- in radical simplifying, 39–42
- trinomial, 133–134

Perpendicular lines, 181–182

Pi (π), 6, 65, 66

Plus sign (+), 8, 70, 83

Points
- choice of, 193
- coordinates of. *See* Coordinates
- determining, 173, 173f

diagram of, 172f
distance between two, 176–177
equal, 173–174
a line through two
parallel line to, 181
perpendicular line to, 181–182
point-slope equation for, 190–193
slope of, 178–181
slope y-intercept equation for, 184–190
midpoint between two, 177
in quadrants, 174–175
from x-y T-table, 185–186
Polynomials
addition of, 90–91
binomials. *See* Binomials
definition of, 85
dividing, 110–117, 120
factoring, 119–137
identifying, 86
monomials. *See* Monomials
multiplying, 94–102
in polynomial expressions, 104–105
in rational expressions, 139
simplifying, 91, 119–120
subtraction of, 92–93
trinomials. *See* Trinomials
Polynomial expressions, 104–109
Positive integer, 3, 3f
Positive numbers
absolute value of, 16–17
addition of, 17–19
categorization of, 7
cube of, 36, 45
cube root of, 7, 36–39
in division, 29–30
exponentiation of
with fractions, 53–57
with natural numbers, 44–48
with negative numbers, 50–52
with one, 45
with zero, 49
fourth root of, 38
integers, 3, 3f
multiplication of, 26–28
natural. *See* Natural numbers
nth root of, 38
on number line, 3–4, 3f, 6, 160f
sign for, 3
square of, 32, 45
square roots of, 5, 7, 32–35, 39–42
subtraction of, 21–24
whole. *See* Whole numbers
Power of a difference rule, 80, 81, 220
Power of a product rule, 78–80, 119–120
Power of a quotient rule, 79
Power of a sum rule, 80, 81, 119–120, 220
Power to a power rule, 77–78
Prime factor, 148
Principal cube root, 36–37, 39, 53
Principal nth root of a, 38–39
Principal square root, 32–36, 38, 53, 164
Product rule, 74–75, 81

Quadratic equations, 127–133, 163–169
Quadratic function, 201, 202f
Quotient rule, 75–77, 81

Radicals, 38–42
Radicand, 38
Range of functions, 195–196, 197
Rational expressions
algebraic fractions. *See* Algebraic fractions
definition of, 139
fundamental principles of, 139–143
Rational numbers
decimal form of, 4–5
definition of, 4
integers, 3–5, 3f, 6f
negative. *See* Negative numbers
nonintegers, 6
positive. *See* Positive numbers
set of, 4
Real numbers
absolute value of, 15–17
addition rules for signed, 17–20
categorization of, 64
comparing, 14–15
definition of, 6
division rules for signed, 29–30
in domain of functions, 196–199
field properties of, 8–12
graph of, 8
irrational, 5–6, 6f
multiplication rules for signed, 26–28
nth root of, 38
on number line, 6, 6f, 8
rational. *See* Rational numbers
subtraction rules for signed, 21–25

Reciprocals
  in dividing of algebraic fractions, 145–146
  of exponential expression, 50–51
  multiplication by, 11
  perpendicular line slopes, 181–182
  simplifying, 51–52
Rise, 178, 178f, 187, 188
Run, 178, 178f, 187, 188

Sets
  correspondence between, 196
  dots in, 1
  function, 195–196
  of integers, 3
  "is an element of" symbol for, 11
  of natural numbers, 1
  of rational numbers, 4
  of whole numbers, 2
Seventh root, 39
Signed numbers. *See* Real numbers
Sixth root, 6, 39
Slope of a line
  definition of, 178
  parallel lines, 181
  perpendicular lines, 181–182
  through two points
    basics, 178–181
    from graph, 178f, 187, 188
    negative, 178, 180, 185
    point-slope equation for, 190–193
    positive, 178–179, 185
    y-intercept equation for, 184–190
  zero, 178, 180–181, 185
Special products
  difference of two cubes, 102, 134, 136
  difference of two squares, 102, 134–135
  perfect cubes, 36, 102
  perfect squares. *See* Perfect squares
  sum of two cubes, 102, 134, 136
Square root
  absolute value and, 32, 36, 164
  categorization of, 5–7
  of decimals, 34, 53
  finding, 32–36
  of fractions, 34–35, 41–42
  function with, 198–200
  in monomials, 85
  of negative numbers, 6, 32, 36, 163
  on number line, 6
  of one, 33
  of positive numbers, 5, 7, 32–35, 39–42
  principal, 32–36, 38, 53, 164
  simplifying, 39–42
  symbol for. *See* Square root symbol
  of zero, 5, 32
Square root function, 202, 202f
Square root symbol
  categorization of, 58
  in order of operations
    with addition, 35, 119
    algebraic expressions, 67–70
    with multiplication, 36, 37, 39–42
    PEMDAS, 60
  sign with, 5, 32
Squares, diagonal length in unit, 5, 5f
Substitution method, 206–208
Subtraction
  of algebraic fractions, 146–150
  commutative property of, 24
  within complex fractions, 151–152
  difference of two cubes, 102, 134, 136
  difference of two squares, 102, 134–135
  of integers, 4
  of like terms, 88–89
  linear equation rules, 155
  linear inequality rules, 160
  of monomials, 88–89
  of natural numbers, 1
  of negative numbers, 21–25
  opposite numbers in, 21–25, 92–93
  in order of operations
    with absolute value, 120
    algebraic expressions, 68–70
    PEMDAS, 58–63
  parentheses in, 147
  of polynomials, 92–93
  power of a difference rule, 80, 81, 220
  in rational expressions, 141–150
  sign for, 21, 25
  signed numbers rules, 21–25
  in slope formula, 178
  of whole numbers, 2
  of zero, 22, 24
Sum of two cubes, 102, 134, 136
Sum of two squares, 134, 135